Application of Sewage Sludge in Industrial Wastewater Treatment

Application of Sewage Sludge in Industrial Wastewater Treatment

Edited By

Maulin P. Shah
Applied & Environmental Microbiology Lab
Gujarat, India

WILEY

This edition first published 2024

Registered Office
John Wiley & Sons, Inc., 111 River Street, Hoboken, NJ 07030, USA

For details of our global editorial offices, customer services, and more information about Wiley products visit us at www.wiley.com.

Wiley also publishes its books in a variety of electronic formats and by print-on-demand. Some content that appears in standard print versions of this book may not be available in other formats.

A catalogue record for this book is available from the Library of Congress

Hardback ISBN: 9781119857365; ePub ISBN: 9781119857389; ePDF ISBN: 9781119857372; oBook ISBN: 9781119857396

Cover Image: © ARTUR PLAWGO/SCIENCE PHOTO LIBRARY/Getty Images
Cover Design: Wiley

Set in 9.5/12.5pt STIXTwoText by Integra Software Services Pvt. Ltd, Pondicherry, India

Contents

1

Sludge Conditioning, Activation, and Engineering

*Joe Cyril Harrish A.M and Hitesh S. Pawar**

DBT-ICT Centre for Energy Biosciences, Institute of Chemical Technology, Matunga, Mumbai, India
** Corresponding author*

1.1 Introduction

During the wastewater treatment process, sewage sludge is generated in massive amounts. It contains a multitude of toxic substances such as heavy metals, organic contaminants, and pathogens. This sewage sludge has severe negative effects in the environment. The water industry is expecting a wide number of constraints due to stringent regulation [1]. In a wastewater treatment plant (WWTP), the cost of the disposal of sewage sludge alone accounts for about 50% [2]. Thus developing a feasible treatment method for this is crucial for waste management. Some of the conventional techniques usually applied for sludge disposal include: composting, incineration, anaerobic digestion, landfill, and recycling. The definition of sewage sludge is that it is the residue produced from the treatment of wastewater. The sludge is comprised of two fractions which are the primary and secondary sludge. The secondary sludge is generated from the biological treatment system. Chemical sludge results from the treatment of sludge with chemicals [3].

Some of the constituents of sewage sludge are protein, fats, nitrogen, phosphoric acid, calcium oxide, cellulose, magnesium oxide, potash, etc. [4]. Sludge components include the chemicals used for treatment. The biochemical transformation during treatment will produce fulvi and humic acids in them. Usually, the treatment methods for wastewater for generating sludge are mechanical treatment such as sedimentation, or sometimes biological process or clarifiers for removal of secondary sludge. The secondary sludge contains harsh chemicals. So, neutralization, coagulation, and precipitation of the above compounds are done. Solids which are removed during mechanical pretreatment are the primary sludge

Application of Sewage Sludge in Industrial Wastewater Treatment, First Edition. Edited by Maulin P. Shah.

which contains a high amount of water, pathogens and they look vile, smell bad and are susceptible, while the secondary sludge is fluffy and yellow in colour, biologically active and dewatering is difficult.

1.2 Conditioning

1.2.1 Conditioning of Sewage Sludge by Increasing the Rate of Hydrolysis

Disintegration is the word for conditioning the sludge by destroying its structure by means of increasing the hydrolysis. Nowadays, more attention is given to the hydrolysis of the sludge. This will in turn increase the biogas production. This biogas can be used as an energy source. Additionally, simultaneous generation of volatile acids during conditioning can be seen. This volatile acid becomes a good carbon source for denitrifiers. The main goal of this type of conditioning is to make the available nutrients accessible to the microbes by disruption of the EPS matrix and cell wall thus aiding hydrolysis (anaerobic digestion) of complex organic molecules of sludge.

1.2.1.1 Mechanical Pretreatment

1.2.1.1.1 Ultrasonic Pretreatment

During ultra-sonication, periodic waves of compression and rarefaction propagate through the wastewater sludge. Microbubbles generated through ultra-sonication collapse rapidly with a time interval of microseconds inducing cavitation. This sudden violent collapse of microbubbles gives rise to extreme temperature of around 5000 K and pressure of around 500 bars. This cavitation generates robust hydro-mechanical shear forces and highly reactive radicals (H · and · OH). The oxidizing effect of reactive radicals and hydro-mechanical shear forces disintegrates the sludge flocs and releases the cellular material [5–7]. The specific energy output of ultra-sonication is optimized based on the total solid content of the wastewater. The main disadvantage of this method is high energy consumption. Specific energy input determines the solubilization efficiency of the sludge. The specific energy input for disintegration of sludge flocs is around 80 KJ L^{-1} [8]. The optimal specific energy input based on total solids was around 1000 KJ Kg^{-1} improves settleability. But for sludge destruction and transformation of insoluble organics into soluble organics, a specific energy input of around 26 000 KJ kg^{-1} of total solids is needed [9]. But this consumes an enormous amount of energy. Ultra-sonication greatly improves the efficiency of anaerobic digestion of sludge. This in turn increases the biogas production (methane). The ultrasound pretreatment method increases the destruction of volatile solids and biogas production by 15–35% [10].

1.2.1.1.2 Microwave Irradiation

Microwave irradiation works in the wavelengths between 1 mm to 1 m and oscillation frequencies in the range 0.3–300 GHz [5, 11]. In theory the damage of sludge flocs can be done in two ways. The first is that dipole rotation under oscillating electromagnetic field generates heat which boils the intracellular fluid and breaks the bacterial cells, and the second way is that changes in dipole orientation of polar molecules breaks the hydrogen bonds and denatures the complex biological molecules [12, 13]. This pretreatment method enhances the solubilization of organic matter in the sludge and increases the biogas production by 50% [14]. This pretreatment method causes rapid lysis of cell residue along with simultaneous disruption of extracellular polymeric substances. The non-thermal effect exerted by disorientation of dipoles has a slight effect on the solubilization of organic matter. But, the mesophilic anaerobic biodegradability of the sludge and biogas production is increased [15]. Although it has its advantages, the microwave irradiation treatment method has no discernible effect on hydrolysis but improves dewaterability and destruction of pathogens responsible for anaerobic digestion.

1.2.1.1.3 Electrokinetic Disintegration

This is a high-voltage electric field method also known as pulsed electric field. During the sludge disintegration process, a high voltage field is induced by charges and causes the rapid disruption of sludge flocs, this makes the nutrients in the effluent easily accessible for the microbes responsible for fermentation [16]. It increases the soluble organic matter by 110–460% in comparison to conventional digestion systems. It decreases the size of the digesters size 40%. The ratio of soluble COD to the total COD increased by 4.5 times and biogas production increased by 2.5 times [17, 18]. The main advantage of this treatment method, is that it drastically reduces the amount of sludge without affecting the treatment efficiency.

1.2.1.1.4 High-pressure Homogenization

During homogenization, the generation of extreme pressure gradient, turbulence, and rapid depressurization creates cavitation and in turn produces strong shearing forces. The sudden pressurization and depressurization breaks the sludge flocs releasing intracellular substances. For sludge disintegration or solubilization, the operating pressure was around 20–80 MPa and the number of cycles was around 1 to 4 [19]. In industry, high pressure homogenization is considered over other pretreatment methods. This is due to ease of operation, high energy-efficiency and low capital cost. This method increases the sludge reduction by 23% and biogas production by 30% [20].

1.2.1.2 Thermal Hydrolysis

Usually, thermal hydrolysis is the commercially accepted and well established pretreatment method for sludge dewaterability [21, 22]. But, research is still underway for its ability to improve sludge disintegration (hydrolysis). The efficiency of this treatment method relies on the treatment temperature and time. The solubilization of sludge increases with increase in operating temperature. The carbohydrates and protein are affected more than lipids in thermal hydrolysis. The operating temperature for sludge solubilization, should not exceed 190 °C for maximum efficiency. Further, increase in temperature decreases the efficiency as it affects the anaerobic digestion. The biodegradability is weakened with further increases in temperature because high temperature results in production of recalcitrant compounds, i.e. melanoidins. These molecules are difficult to degrade and in addition to this even inhibit the biodegradability of organic matter [23]. The treatment time has less influence on solubilization of sludge in comparison to treatment temperature. This technique greatly reduces the hydraulic retention time, this in turn reduces the reactor size for anaerobic digestion. This pretreatment method is advantageous as it destroys pathogens, removs odor, reduces sludge volume, improves dewaterability, etc.

1.2.1.3 Chemical Pretreatment

In this method, the microbial cells in the sludge flocs are deformed by chemicals, this favors enzymatic digestion of organic matter in the sludge. Some of the chemical treatment methods employed are as follows.

1.2.1.3.1 Acidic and Alkali Pretreatment

This methods shows great promise due to ease of operation, high methane conversion efficiency, simple operation, and it is inexpensive [24]. In acid hydrolysis, HCl, H_2SO_4, H_3PO_4, and HNO_3 are used in treatment, while for alkali treatment NaOH, KOH, $Ca(OH)_2$, CaO, and ammonia are used. The major benefit of this method is that it can be performed at ambient temperature [2]. The effectiveness of this method relies on the affinity of the organic matter in the sludge to the acid or alkali and also the characteristics of the sludge. Acid treatment is most suitable for hydrolysis of lignocellulosic biomass but has little to no effect on lignin hydrolysis. The theory behind acid treatment is the hydrolysis of hemicellulose liberating monomeric sugars and oligomers from the matrix of cell wall thereby supporting enzymatic digestibility. The drawback of acid treatment is the generation of toxic byproducts such as furfural, hydroymethyl furufural which strongly inhibits fermentative microbes [25]. Other drawbacks include its corrosive nature due to extremely low pH which can affect the reactor and special measures should be taken into account to prevent the above mentioned problem. To overcome this problem, acidic treatment is integrated with thermal treatment.

On the other hand, alkali treatment is best suitable for the breakdown of lignin. The theory behind this treatment is saponification and solvation, which causes depolymerization and breakage of lignin-carbohydrate linkages [26]. It also saponifies the intermolecular ester bonds of xylan hemicellulose, but the degree of solubilizing them is not as great as acid treatment. The additional alkalinity improves the methanogenic activity and stabilizes the process of anaerobic digestion by acting as a buffer system [27]. NaOH is the most effective in enhancing biogas production and increasing the solubilization of the sludge. Although it has its advantages, some of the disadvantages such as NaOH treatment is not compatible with anaerobic digestion. As overloading NaOH blocks the metabolic pathway of anaerobic microbes and decreases the methane productivity. To overcome this problem, alkali treatment is integrated with microwave, ultrasound, and thermal pretreatment systems in order to reduce the alkali consumption and this also increases the methane productivity.

1.2.1.3.2 Ozonation

This is the most widely employed peroxidation process as it disrupts cell membrane and disintegrates the sludge flocs resulting in excessive solubilization of sludge. Ozonation is integrated with anaerobic sludge digestion to bypass hydrolysis and increase methane productivity [28, 29].

The efficiency of sludge solubilization depends on the dosage and amount of ozone introduced in a moderate range [30]. However, ozonation efficiency depends on the reaction kinetics and mass transfer of ozone. The rate of kinetic reaction between sludge mixed liquor and dissolved ozone is average, even though the applied dosage is high [31]. The introduction of a high dosage of ozone can cause complete mineralization of liberated cellular components affecting methane productivity [32]. The main disadvantage of this method is the high capital investment

The microbubble ozonation was introduced to overcome the high capital investment with higher performance. The sludge solubilization was increased from 15–30% to 25–40% with the bubble contactor with ozone doses of 0.06–0.16g $O_3\ g^{-1}$ of total solids and ozone utilization also improved from 72 to 99%. In this process, rapid generation of hydroxyl radicals is seen thereby accelerating the solubilization efficiency of sludge [33].

1.2.1.3.3 Fenton Oxidation

This is a catalytic reaction involving reaction between hydrogen peroxide and the catalyst iron ions(Fe^{2+}). The result generates enormous amounts of highly active radicals ($\cdot$OH). The hydroxyl radicals generated from this reaction have high oxidative potential of + 2.80 V which is greater than the hydroxyl radicals produced from ozonation and hydrogen peroxide alone. These are effective for the disintegration of sludge flocs (EPS) and disruption of cells [22, 34]. This releases the bound water and intracellular material. Release of bound water enhances dewaterability. In addition to this, the Fenton process enhances methane productivity and sludge solubilization. The efficiency of this process depends on many variables such as treatment time, intial pH and temperature, Fe^{2+}/H_2O_2, and reagent concentration [22, 35]. This method shows greater efficiency by increasing the soluble COD by five times and methane productivity increased by 75% [7]. The major drawback on Fenton oxidation is that it needs extremely low pH to prevent precipitation and hydrolysis of Fe^{3+} [2, 36, 37]. Since for Fenton oxidation, enormous amounts of H_2O_2 and Fe^{2+} will be present. This will scavenge the hydroxyl radicals.

1.2.1.3.4 Fe(II)-activated Persulfate Oxidation

This is a new and emerging sludge pretreatment technology for efficient dewatering of sludge. In this method, persulfate ($S_2O_8^{2-}$) molecules will be activated by heat and UV light or by transition metals. This will liberate an extremely strong oxidant which is sulfate free radicals (SO_4^{-}). These free radicals have a redox potential of around 2.60 V. These free radicals oxidize and attack EPS and bacterial cells and particularly target aromatic peptides, tryptophan protein, humic, and fulvic acid-like substances of EPS. Then the simultaneous cleavage of the polymeric back-bone of cells results in the rupture of their cells. The major advantage of this method over Fenton oxidation is that sulfate radicals have higher oxidation potential over wide range of pH which makes them cost effective [38].

1.2.2 Conditioning for Sludge Dewatering

Sludge dewatering decreases the sludge volume for ease of transportation, energy utilization, and reduction of leachate production in landfill sites on disposal. Some of the conditioning technologies are as follows (Table 1.1).

1.2.2.1 Physical Conditioning

1.2.2.1.1 Porous Material Addition

Sludge dewaterability effectiveness is assessed based on minimizing the compressibility and increasing the permeability of sludge. Some of the porous materials include coal fly ash, wood chips, gypsum, and lignite and rice husk. The addition of porous material to the wastewater in the proportion of 30–50% dry matter can be followed in the wastewater treatment. This treatment method reduces the HRF by 58–88% and compressibility coefficient by 22–90%. Inorganic porous material has an advantage over organic porous material as it reduces the phosphorus content after filtration by the formation of insoluble phosphate. This lessens the negative charges on the sludge and improves sludge flocculation. Mineral based adsorbent has a disadvantage over carbon based adsorbent as it has

Table 1.1 Conditioning for increasing the rate of hydrolysis.

s.no	Pretreatment	Condition	Performance	Reference
1	Sonication of activated sludge	3380 KJ kg^{-1}	42% methane yield, 13% voltalile solid removal	[39]
2	Sonication of mixed sludge	5000 KJ kg^{-1}	35% methane yield	[40]
3	Microwave irradiation of activated sludge	14 000 KJ kg^{-1}	570.7% biogas production	[41]
4	Microwave irradiation of thickened sludge	-	106% biogas production, 53.1% volatile solid removal	[42]
5	Electrokinetic disintegration of activated sludge	10 kWh m^{-2}	100% methane production	[43]
6	Electrokinetic disintegration of primary sludge	33 kWh m^{-3}	33% methane production	[44]
7	High pressure homogenization of mixed sludge	12000 psi	0.61–1.32 of methane/Ld 43–64% of volatile solids removal	[45]
8	High pressure homogenization of concentrated sludge	150 bar	30% gas production	[46]
9	Thermal hydrolysis of secondary sludge	134–140 °C 3.4 bar, 30 min	Volatile removal 42%	[47]

lower capabilities of improving the permeability of the sludge in comparison. Also carbon based materials do not decrease the calorific value of the sludge. The porous material should not increase the volume of the dewatered sludge and porous material particle size should be more than 10 μm in order to not block the filter medium [48].

1.2.2.1.2 Sonication

Low frequency ultrasound generates acoustic cavitation through mechanical shear forces. This disrupts the cell membrane and liberates intracellular organics into the aqueous medium. This helps in disintegration of the biological flocs and homogenization of sludge thus dewatering of the sludge happens. The homogenized sludge has sponge-like properties. Although, it is debatable based on studies, ultra-sonication may have negative effects on dewaterability [49–51].

Hydrodynamic cavitation is another process using the venturi or orifice plate in the flowpath of liquid which in turn increases the velocity of the flow during reduction of the cross-sectional area. This increase in velocity decrease the local pressure resulting in cavitation. The cavitation produced by this is similar to ultra sonication. The critical parameter in the hydrodynamic cavitation is the optimization of flow regime for positive effects [52].

1.2.2.1.3 Freezing and Thawing

Freezing/thawing is an effective way of sludge dewatering. During this process, sludge is first frozen at a temperature below abpit −20 to −15 °C which results in the formation of ice crystals and this is followed by thawing at room temperature. Sludge particles were excluded from the ice crystals and water molecules were trapped in the ice crystals. While thawing, the concentrated solid was segregated from the water. The continuous freezing/thawing cycle causes crystallization of intracellular water and its volume expands. It is accompanied by swelling of cells and lessens the bound water content which will result in heavy sludge dehydration. The only limiting factor in implementation of this technology is high energy consumption [52–54]

1.2.2.1.4 Electrical Treatment

The integration of mechanical compression with electrical treatment enhanced the sludge dewatering and reduces the water content by up to 50%. The sludge particles usually have amphoteric groups which makes them negatively charged. This drives them towards the cathode/anode based on their charges. The concentration gradient of ions in the water phase is created by electrically driven movement of hydrated ions and enhances the water diffusion in sludge. In the permeable polar plate, water is removed from the sludge through water osmosis. Other ways of removal in this process include electrolysis of water. These processes further enhance the dewatering of sludge [55].

1.2.2.2 Chemical Conditioning

By altering the physico-chemical properties of the sludge by the use of chemicals for improving sludge dewaterability is referred gto as chemical conditioning. Some types of chemical conditioning methods are enzymatic treatment, coagulation/flocculation, advanced oxidation process, and acid/base treatment (Table 1.2).

1.2.2.2.1 Coagulation/flocculation

Aluminium and ferric salts are the most commonly used inorganic salt coagulants. In comparison, aluminium salts are more effective than ferric salts which result in rapid settling. However, due to low cost, rapid settling velocity, flexible pH, lower toxicity, and better effects at low temperature ferric salts are preferred

Table 1.2 Conditioning for dewatering of sludge.

s.no	Pretreatment	Hypothesis	limitation	Reference
1	Porous material addition	van der Waals forces	Disposal cost	[63]
2	Sonication	Disintegration of bioflocs, cell rupture for releasing of water	Specialized equipment, high capital cost	[63]
3	Thermal treatment	Bioflocs disintegration/ rupturing of cell for releasing water	High capital cost, difficulty to scale up	[63]
4	Freezing/ thawing	Homogenous ice crystal formation and cell swells to release moisture	High capital cost, difficulty to scale up	[63]
5	Electrical treatment	Electromigration, electrophoresis, electro osmosis, electrolysis,	High capital cost, specialized walls for anticorrosion	[63]
6	Acid/base treatment	Regulation of surface charge and cell rupture	cororsiveness	[63]
7	Advanced oxidation process	Lysis of EPS and cell rupture	Robust pH control	[63]
8	Enzymatic treatment	EPS degradation	Harsh conditions and long duration process	[63]

over aluminium salts. Some of the disadvantages of aluminium salts include high consumption as most of the aluminium ions will be consumed by soluble-EPS for chelation and for protein-like substances ferric ions show more affinity than aluminium ions. The drawback of using ferric salts is that it may corrode the walls of the equipment [56, 57].

The polymerized form of aluminum ions show more efficiency in coagulation in comparison to the monomeric form of aluminium as it forms sludge flocs with small particles with low compressibility resulting in denser flocs. This is due to the fact that multivalent inorganic compounds have charge density and large molecular mass and this could be the reason for balance of charge even at low dosage [58].

The adsorption-bridging reaction on the particles of sludge is the principle behind the macromolecular polymers known as flocculants. This improves the solid–liquid separation by creating a density difference between the flocs and the liquid. Almost all the sludge particles are negatively charged, cationic polyacrylamide is the most commonly used flocculant for dewatering of sludge. They are effective in destabilization of sludge flocs through neutralization of charge and re-aggregates the sludge particles [59, 60].

1.2.2.2.2 Advanced Oxidation Process

The bound water content is due to the presence of EPS in the sludge. The dewaterability improvement of the sludge relies on the fact that AOP can destroy the EPS of the sludge flocs. The non-selective and highly reactive · OH facilitates the depolymerization and destruction of EPS [61]. Some of the chemicals employed in AOP are $Na_2S_2O_8$, $NaHSO_5$, and these can perform well under neutral pH range. In recent years, to preventing high dosage of conditioning agents, heterogeneous Fenton reaction was applied to the sludge flocs as it uses zero valence iron as catalyst. The extracellular protein content in the cells are mostly hydrophilic components which determine the influence in the dewaterability of the sludge rather than the humic acids and polysaccharides present in the sludge biomass [62].

1.2.2.2.3 Enzymatic Treatment

Some of the enzymes used in sludge dewatering are proteinase, DNase, polygalacturonase, cellulase, and amylase. These enzymes can induce the EPS degradation which can limit the water holding capacity of the sludge. The main advantage of this method over AOP is the dosage, as the sludge dewaterability can be done in low concentration. The drawbacks of using enzymatic treatment is robust reaction conditions and long retention time [63].

1.3 Activation of Sewage Sludge

Since the disposal of sludge has many limitations such as cost issues during incineration and liberation of huge quantities of hazardous ash, and landfill disposal has many environmental after-affects. By addressing this problem, a suitable solution is to valorize them by converting them into adsorbents especially activated carbon. Initially, sludges which are rich in carbon are heated at a temperature around 500–800 °C for carbonization. This process will eliminate the non-carbon fraction through the release of gaseous volatile products. The decomposition of these sludge particles at high temperature will increase porousity with abundant surface area. The structure of the pyrolyzed sludge is as graphite crysrallites [64, 65].

Although its been pyrolyzed, some of the pores will be blocked due to disorganized carbon or tarry residues. To make them more efficient as adsorbents there are activation methods such as physical activation and chemical activation (Table 1.3). Although self-activation also comes under the above category.

Table 1.3 Activation of sewage sludge through different activation.

S.no	Sludge	Carbonization condition	Activation condition	BET surface area	Reference
1	Sludge from wastewater treatment plant	No carbonization	Steam 700 °C	55	[73]
2	Municipal sludge WWTP sludge cake	700 °C, 1 hr	Steam 800 °C	88	[74]
3	Dewatered municipal aerobically digested activated sludge WWTP	700 °C, 0.5 hr	CO_2, 800 °C, 2–4 hr	269	[75]
4	Municipal WWTP, viscous liquid cake	600°C, 1hr	Steam, 760°C, 0.5	226	[76]
5	Anerobically digested dewatered municipal sludge	Not carbonized	10g in 25ml, 5M $ZnCl_2$,500°C, 2hr	647	[77]
6	Municipal waste	Not carbonized	3M H_2SO_4, 100g in 250ml 3M, 650°C	408	[78]

1.3.1 Adsorbents Produced by Thermal Carbonization

The Brunauer, Emmett, and Teller (BET) surface area of the carbonization is highly influenced by both temperature and carbonization time, inorganic content, moisture content, and composition of the sewage sludge The surface area of the sludge will increase with an increase in temperature due to enhanced aromatization at high temperature and will vaporize water through dehydroxylation of the inorganic material. This increase in surface area will increase the pores and dehydroxylation will be the activation agent for microspores. A too high temperature will result in shrinkage and solidification of the carbonaceous matrix. The reason behind this is the surfeit aromatization which will narrow or even closes the pores. This will lessen the accessible surface area [66–68]. The pores are categorized as macrospores, mesoporous, and microspores based on the pore size. The diversity of the pores is influenced by the presence of microorganisms in the sludge especially their cell walls. The cell wall comprises of peptidoglycans and polysaccharides. Comparatively, polysaccharides are volatized at low temperature thus forming mesoporous and macrospores. Macrospores are usually formed due to intermediate melt in the sludge, as it softens and bubbles out as gas and other ways including moisture vaporization and volatilization of low molecular weight volatile molecules [69].

Microspore formation can be improved by pre-drying. During pre-drying, oxygen–carbon interaction occurs resulting in thermosetting of the sludge. This will inhibit the formation of intermediate products [66].

Other factors that influence the surface area of the adsorbents include humic acid to fulvic acid ratio. The higher the fulvic acid the higher the surface area it influences as it facilitates the release of oxygen and hydrogen [70]. Sewage sludge blends have high surface areas, some illustration include blending of sewage sludge with commercial activated carbon, with sulphonic acid- co-maleic acid, sodium salt of polystyrene. HCL washing of sewage sludge increases the BET surface area more than 10 times. The increase can be attributed to inorganic matter removal through the above washing, but at the same time increase the accessibility to carbon fraction.

1.3.2 Adsorbents by Physical Activation

The physical inactivation will be initiated by burning off the carbon fraction of the sludge by gasification. The inorganic content after physical activation is quite high. It will be more than 60% of the total weight of the sludge based adsorbent [71]. Although the surface area is very low, the ratio of microspore to mesoporous is high. For gasification, steam is normally used. But CO_2 is favorable in comparison as it possesses lower reactivity to steam.

Since the concentration of inorganic constituents is very high, this will probably reduce the BET surface area of the adsorbent. So, to tackle this acid washing especially HCl washing was employed to increase the surface area of the adsorbent. Washing of adsorbent before activation will be effective in comparison to post washing. The surface area of the adsorbent was increased by nearly 100 times after washing [72].

1.3.3 Adsorbents after Chemical Activation

Some of the commonly used chemicals in this activation include H_3PO_4, K_2S, H_2SO_4, KOH, $ZnCl_2$, and NaOH. The principle behind the activation of adsorbent with bases is that it reacts directly with the carbon atoms to improving the porosity. Usually, after carbonization the ratio of C:H and C:O will be very high. This will be an added advantage for effective impregnation of KOH. Dry mixing of KOH with carbonized adsorbent is more efficient than wet impregnation. In KOH activation, inorganic fraction content shows positive effects by increasing the surface area. This can be explained by the formation of amorphous silica and alkaline fusion which makes the adsorbent highly porous. Apart from this, NaOH and $ZnCl_2$ are the second and third most used chemicals after NaOH [72].

$ZnCl_2$ improves the carbon skeleton aromatization simultaneously improving pore structure and it will be developed during volatilization. Heating rate and boiling point of $ZnCl_2$ is a limiting factor. Above boiling point, volatilization of the salt occurs and at low heating rate there will be no deformation or collapse during the release of gas and development of the pore [72]. An acid such as H_2SO_4 is preferred over $ZnCl_2$ due to its low environmental impact even though the efficiency is low. The main advantage of using sulphuric acid is that it can be used at low temperatures.

Some other activation agents include phosphoric acid, K_2S. Today, even a combination of physical and chemical activation has been performed.

1.3.4 Surface Chemistry of the Sewage Sludge-Based Adsorbent

High porosity is a characteristic for a good adsorbent. However, if the surfaces of the adsorbent do not have any affinity for the target molecule then it is not for that compound. So, surface chemistry of the adsorbent is needed more in comparison to high porosity. Some of the important parameters to be taken into consideration for surface chemistry are pH, surface functional groups, and specific adsorbate uptake per unit area [72].

The pH range of alkaline sewage sludge adsorbents will be around 11.4. During carbonization, polycondensation and degenerative polymerization, increased inorganic content and the most important reason will be the destruction of acidic functional groups which will raise the pH. Some other additional causes may be due to the destruction of inorganic salts. By using additives, the pH can be increased by addition of calcium oxide after stabilization of the carbonization of the sludge. Sludge based adsorbents producing physical activation usually shows higher basicity than just carbonized sludge. Usually HCl washing decreases the pH of the sludge based adsorbents below 4.5, when $ZnCl_2$ activated adsorbent has pH below 2.5 without washing. Usually, sludge based adsorbents have an average content of phenol and carboxylic groups, hydrophilicity and some amount of cation exchange ability. Sulphuric acid activation produces sludge based adsorbent with acidic surface. Other techniques for lowering pH can be treatment using oxygen containing surface functional groups [72].

1.4 Conclusion

Both industrial and municipal wastewater treatment plants generate sewage sludge in large amounts and they are increasing day by day. This is due to population explosion and rapid establishment of industries all over the world. So, there must be some robust sludge treatment techniques and efficient valorization of the sludge must be established. This will limit the pollution to a greater extent and at the same time be cost effective.

References

1 Park, N.D., Helle, S.S., and Thring, R.W. (2012). Combined alkaline and ultrasound pre-treatment of thickened pulp mill waste activated sludge for improved anaerobic digestion. *Biomass & Bioenergy* 46: 750–756.

2 Appels, L., Baeyens, J., Degreve, J., and Dewil, R. (2008). Principles and potential of the anaerobic digestion of waste-activated sludge. *Progress in Energy and Combustion Science* 34: 755–781.

3 Ren, S. (2004). Assessing wastewater toxicity to activated sludge: recent research and developments. *Environment International* 30 (8): 1151–1164.

4 Turovskiy, I.S. and Mathai, P.K. (2006). *Wastewater Sludge Processing*. Hoboken, New Jersey: John Wiley & Sons, Inc.

5 Harris, P.W. and McCabe, B.K. (2015). Review of pre-treatments used in anaerobic digestion and their potential application in high-fat cattle slaughterhouse wastewater. *Applied Energy* 155: 560–575.

6 Tiehm, A., Nickel, K., Zellhorn, M., and Neis, U. (2001). Ultrasonic waste activated sludge disintegration for improving anaerobic stabilization. *Water Research* 35: 2003–2009.

7 Chu, C.P., Lee, D.J., Chang, B.V., You, C.S., and Tay, J.H. (2002). "Weak" ultrasonic pre-treatment on anaerobic digestion of flocculated activated biosolids. *Water Research* 36: 2681–2688.

8 Foladori, P., Laura, B., Gianni, A., and Giuliano, Z. (2007). Effects of sonication on bacteria viability in wastewater treatment plants evaluated by flow cytometry – fecal indicators, wastewater and activated sludge. *Water Research* 41: 235–243.

9 Feng, X., Lei, H., Deng, J., Yu, Q., and Li, H. (2009). Physical and chemical characteristics of waste activated sludge treated ultrasonically. *Chemical Engineering and Processing* 48: 187–194.

10 Long, J.H. and Bullard, C.M. (2014). Waste activated sludge pretreatment to boost volatile solids reduction and digester gas production: market and technology assessment. *Florida's Water Resources Journal* 44–50.

11 Tyagi, V.K. and Lo, S.L. (2013). Microwave irradiation: a sustainable way for sludge treatment and resource recovery. *Renewable and Sustainable Energy Reviews* 18: 288–305.

12 Tang, B., Yu, L., Huang, S., Luo, J., and Zhuo, Y. (2010). Energy efficiency of pre-treating excess sewage sludge with microwave irradiation. *Bioresource Technology* 101: 5092–5097.

13 Toreci, I., Kennedy, K.J., and Droste, R.L. (2009). Evaluation of continuous mesophilic anaerobic sludge digestion after high temperature microwave pretreatment. *Water Research* 43: 1273–1284.

14 Appels, L., Houtmeyers, S., Degreve, J., Van Impe, J., and Dewil, R. (2013). Influence of microwave pre-treatment on sludge solubilization and pilot scale semi-continuous anaerobic digestion. *Bioresource Technology* 128: 598–603.

15 Eskicioglu, C., Terzian, N., Kennedy, K.J., Droste, R.L., and Hamoda, M. (2007). Athermal microwave effects for enhancing digestibility of waste activated sludge. *Water Research* 41: 2457–2466.

16 Krzemieniewski, M., Zieliński, M., and Dębowski, M. (2016). Technology of electrokinetic disintegration of virginia fanpetals (sidahermaphrodita) biomass in a biogas production system. *International Journal of Energy and Power Engineering* 3: 759.

17 Lee, I.S. and Rittmann, B.E. (2011). Effect of low solids retention time and focused pulsed pretreatment on anaerobic digestion of waste activated sludge. *Bioresource Technology* 102: 2542–2548.

18 Choi, H., Jeong, S.W., and Chung, Y.J. (2006). Enhanced anaerobic gas production of waste activated sludge pretreated by pulse power technique. *Bioresource Technology* 97: 198–203.

19 Zhang, Y.X., Zhang, P.Y., Guo, J.B. et al. (2013). Sewage sludge solubilization by high-pressure homogenization. *Water Science and Technology : A Journal of the International Association on Water Pollution Research* 67: 2399–2405.

20 Onyeche, T.I. (2007). Economic benefits of low pressure sludge homogenization for wastewater treatment plants. In: *IWA Specialist Conferences Moving Forward Wastewater Biosolids Sustainability*, 417–422. Moncton, New Brunswick, Canada.

21 Haug, R.T., Stuckey, D.C., Gossett, J.M., and McCarty, P.L. (1978). Effect of thermal pretreatment on digestibility and dewaterability of organic sludges. *Journal (Water Pollution Control Federation)* 50: 73–85.

22 Neyens, E. and Baeyens, J. (2003). A review of thermal sludge pre-treatment processes to improve dewaterability. *Journal of Hazardous Materials* 98: 51–67.

23 Climent, M., Ferrer, I., Baeza, M.D. et al. (2007). Effects of thermal and mechanical pretreatments of secondary sludge on biogas production under thermophilic conditions. *The Chemical Engineering Journal* 133: 335–342.

24 Zhen, G.Y., Lu, X.Q., Li, Y.Y., and Zhao, Y.C. (2014). Combined electrical-alkali pretreatment to increase the anaerobic hydrolysis rate of waste activated sludge during anaerobic digestion. *Applied Energy* 128: 93–102.

25 Harmsen, P.F.H., Huijgen, W.J.J., LMB, L., and Bakker, R.R.C. (2010). *Literature review of physical and chemical pre-treatment processes for lignocellulosic biomass.* A review report published by Wageningen UR Food & $2 Biobased Research. NL6700 AA.

26 Ariunbaatar, J., Panico, A., Esposito, G., Pirozzi, F., and Lens, P.N.L. (2014). Pretreatment methods to enhance anaerobic digestion of organic solid waste. *Applied Energy* 123: 143–156.

27 Harmsen, P.F.H., Huijgen, W.J.J., LMB, L., and Bakker, R.R.C. *Literature review of physical and chemical pre-treatment processes for lignocellulosic biomass.* A review report published by Wageningen UR Food & $2 Biobased Research. NL6700 AA; 20.

28 Sakai, Y., Fukase, T., Yasui, H., and Shibata, M. (1997). An activated sludge process without excess sludge production. *Water Science and Technology : A Journal of the International Association on Water Pollution Research* 36: 163–170.

29 Campos, J.L., Otero, L., Franco, A., Mosquera-Corral, A., and Roca, E. (2009). Ozonation strategies to reduce sludge production of a seafood industry WWTP. *Bioresource Technology* 100: 1069–1073.

30 Bougrier, C., Battimelli, A., Delgenes, J.P., and Carrere, H. (2007). Combined ozone pretreatment and anaerobic digestion for the reduction of biological sludge production in wastewater treatment. *Ozone: Science and Engineering* 29: 201–206.

31 Manterola, G., Uriarte, I., and Sancho, L. (2008). The effect of operational parameters of the process of sludge ozonation on the solubilization of organic and nitrogenous compounds. *Water Research* 42: 3191–3197.

32 Carrere, H., Dumas, C., Battimelli, A. et al. (2010). Pretreatment methods to improve sludge anaerobic degradability: a review. *Journal of Hazardous Materials* 183: 1–15.

33 Chu, L.B., Yan, S.T., Xing, X.H. et al. (2008). Enhanced sludge solubilization by microbubble ozonation. *Chemosphere* 72: 205–212.

34 Pilli, S., Yan, S., Tyagi, R.D., and Surampalli, R.Y. (2015). Overview of Fenton pre-treatment of sludge aiming to enhance anaerobic digestion. *Reviews in Environmental Science and Biotechnology* 14: 453–472.

35 Erden, G. and Filibeli, A. (2010). Improving anaerobic biodegradability of biological sludges by Fenton pre-treatment: effects on single stage and two-stage anaerobic digestion. *Desalination* 251: 58–63.

36 Badawy, M.I. and Ali, M.E.M. (2006). Fenton's peroxidation and coagulation processes for the treatment of combined industrial and domestic wastewater. *Journal of Hazardous Materials* 136: 961–966.

37 Zhen, G.Y., Lu, X.Q., Wang, B.Y. et al. (2014). Enhanced dewatering characteristics of waste activated sludge with Fenton pretreatment: effectiveness and statistical optimization. *Frontiers of Environmental Science & Engineering* 8: 267–276.

38 Zhen, G.Y., Lu, X.Q., Li, Y.Y. et al. (2012). Novel insights into enhanced dewaterability of waste activated sludge by Fe(II)-activated persulfate oxidation. *Bioresource Technology* 119: 7–14.

39 Riau, V., De la Rubia, M.A., and Perez, M. (2015). Upgrading the temperature-phased anaerobic digestion of waste activated sludge by ultrasonic pretreatment. *The Chemical Engineering Journal* 259: 672–681.

40 Martin, M.A., Gonzalez, I., Serrano, A., and Siles, J.A. (2015). Evaluation of the improvement of sonication pre-treatment in the anaerobic digestion of sewage sludge. *The Journal of Environmental Management* 147: 330–337.

41 Ebenezer, A.V., Arulazhagan, P., Kumar, S.A., Yeom, I.T., and Banu, J.R. (2015). Effect of deflocculation on the efficiency of low-energy microwave pretreatment and anaerobic biodegradation of waste activated sludge. *Applied Energy* 145: 104–110.

42 Houtmeyers, S., Degreve, J., Willems, K., Dewil, R., and Appels, L. (2014). Comparing the influence of low power ultrasonic and microwave pre-treatments on the solubilization and semi-continuous anaerobic digestion of waste activated sludge. *Bioresource Technology* 171: 44–49.

43 Salerno, M.B., Lee, H.S., Parameswaran, P., and Rittmann, B.E. (2009). Using a pulsed electric field as a pretreatment for improved biosolids digestion and methanogenesis. *Water Environment Research: A Research Publication of the Water Environment Federation* 81: 831–839.

44 Ki, D., Parameswaran, P., Popat, S.C., Rittmann, B.E., and Torres, C.I. (2015). Effects of prefermentation and pulsed-electric-field treatment of primary sludge in microbial electrochemical cells. *Bioresource Technology* 195: 83–88.

45 Wahidunnabi, A.K. and Eskicioglu, C. (2014). High pressure homogenization and two-phased anaerobic digestion for enhanced biogas conversion from municipal waste sludge. *Water Research* 63: 430–446.

46 Zhang, S., Zhang, P.Y., Zhang, G.M., Fan, J., and Zhang, Y.X.(2012). Enhancement of anaerobic sludge digestion by high-pressure homogenization. *Bioresource Technology* 118: 496–501.

47 Qi, Y., Thapa, K.B., and Hoadley, A.F. (2011a). Application of filtration aids for improving sludge dewatering propertiesea review. *The Chemical Engineering Journal* 171 (2): 373–384.

48 Chen, Y., Yang, H., and Gu, G. (2001). Effect of acid and surfactant treatment on activated sludge dewatering and settling. *Water Research* 35 (11): 2615–2620.

49 Gogate, P.R. and Kabadi, A.M. (2009). A review of applications of cavitation in biochemical engineering/ biotechnology. *Biochemical Engineering Journal* 44 (1): 60–72.

50 Feng, X., Deng, J., Lei, H. et al. (2009b). Dewaterability of waste activated sludge with ultrasound conditioning. *Bioresource Technology* 100 (3): 1074–1081.

51 Mowla, D., Tran, H., and Allen, D.G. (2013). A review of the properties of biosludge and its relevance to enhanced dewatering process. *Biomass & Bioenergy* 58 (1): 365–378.

52 Martel, C.J. (1993). Fundamentals of sludge dewatering in freezing beds. *Water Science and Technology* 28 (1): 29–35.

53 Tuncal, T. (2011). Improving thermal dewatering characteristics of mechanically dewatered sludge: response surfsce analysis of combined lime-heat treatment. *Water Environment Research* 83 (5): 405–410.

54 Mahmoud, A., Hoadley, A.F., Citeau, M. et al., 2018b. *A comparative study of electro-dewatering process performance for activated and digested wastewater sludge*,129, 66–82.

55 Turchiuli, C. and Fargues, C. (2004). Influence of structural properties of alum and ferric flocs on sludge dewaterability. *The Chemical Engineering Journal* 103 (1–3): 123–131.

56 Yang, P., Li, D., Zhang, W. et al. (2019). Flocculation –dewatering behaviour of waste activated sludge particles under chemical conditioning with inorganic polymer flocculant;effects of typical sludge properties. *Chemosphere* 218: 930–940.

57 Zhao, M., Wang, F., Fan, Y., Raheem, A., and Zhou, H.(2019). Low temperature alkaline pyrolysis of sewage sludge for enhanced H2 production with in-situ carbon capture. *International Journal of Hydrogen Energy* 44 (16): 8020–8027.

58 Yang, Z.L., Gao, B.Y., Li, C.X., Yue, Q.Y., and Liu, B. (2010). Synthesis and characterization of hydrophobically associating cationic polyacrylamide. *The Chemical Engineering Journal* 161 (1e2): 27–33.

59 Sun, Y., Zhu, C., Xu, Y. et al. (2016). Comparison of intiation methods in the structure of CPAM and sludge flocs properties. *Journal of Applied Polymer Science* 33 (40): 44–52.

60 Buyukkamaci, N. (2004). Biological sludge condition by Fentons reagent. *Process Biochemistry* 39 (11): 1503–1506.

61 Li, Y., Yuan, X., Wang, D. et al. (2018). Recyclabale zero-valent iron activating peroxymonosulfate synchronousslycombined with thermal treatment enhances sludge dewaterability of digested sludge and increasing biogas production through thermochemical hydrolysis process. *Proceedings of the Water Environment Federation* 2017 (5): 5062–5069.

62 Lu, J., Rao, S., Le, T., Mora, S., and Banerjee, S. (2011). Increasing cake solids of cellulosic sludge through enzyme assisted dewatering. *Process Biochemistry* 46 (1): 353–357.

63 Wu, B., Dai, X., and Chai, X. (2020). Critical review on dewatering of sewage sludge: influential mechanism, conditioning technologies and implications to sludge re-utilizations. *Water Research* 180: 115912.

64 Bansal, R.C. and Goyal, M. (2005). *Activated Carbon Adsorption*. Boca Raton, FL: Taylor & Francis.

65 Smisek, M. (1967). Manufacture of active carbon. In: *Active Carbon: Manufacture, Properties and Applications* (ed. M. Smisek and S. Cerny), 10–27. Barking, Essex, England: Elsevier Publishing Co., Ltd.

66 Lu, G.Q., Low, J.C.F., Liu, C.Y., and Lua, A.C. (1995). Surface-area development of sewage-sludge during pyrolysis. *Fuel* 74 (3): 344–348.

67 Zhai, Y.B., Wei, X.X., and Zeng, G.M. (2004a). Effect of pyrolysis temperature and hold time on the characteristic parameters of adsorbent derived from sewage sludge. *Journal of Environmental Sciences-China* 16 (4): 683–686.

68 Inguanzo, M., Menendez, J.A., Fuente, E., and Pis, J.J. (2001). Reactivity of pyrolyzed sewage sludge in air and CO2. *Journal of Analytical and Applied Pyrolysis* 58–59: 943–954.

69 Talaro, K.P. and Talaro, A. (1999). *Foundations in Microbiology: Basic Principles*. Boston: McGraw-Hill College.

70 Gasco, G., Blanco, C.G., Guerrero, F., and Mendez Lazaro, A.M. (2005a). The influence of organic matter on sewage sludge pyrolysis. *Journal of Analytical and Applied Pyrolysis* 74 (1–2): 413–420.

71 Nickerson, R.D. and Messman, H.C., 1975. *Making active carbons from sewage sludge*. Combustion Engineering Inc. US Patent Office, Patent No. 3,887,461.

72 Smith, K.M., Fowler, G.D., Pulket, S., and Graham, N.J.D. (2009). Sewage sludge-based adsorbents: a review of their production, properties and use in water treatment application. *Water Research* 43: 2569–2594.

73 Bilitewski, B. (1979). *Production and possible applications of activated carbon from waste*. In: (ed. K.J. Thome-Kozmiensky). Berlin: E. Freitag – Verlag Umwelttechnik. 714–721.

74 Fitzmorris, K.B., Lima, I.M., Marshall, W.E., and Reimers, R.S. (2006b). Anion and cation removal from solution using activated carbons from municipal sludge and poultry manure. *Journal of Residuals Science and Technology* 3 (3): 161–167.

75 Ros, A., Lillo-Rodenas, M.A., Fuente, E. et al. (2006a). High surface area materials prepared from sewage sludge-based precursors. *Chemosphere* 65 (1): 132–140.

76 Rio, S., Le Coq, L., Faur, C., and Le Cloirec, P. (2006a). Production of porous carbonaceous adsorbent from physical activation of sewage sludge: application to wastewater treatment. *Water Science and Technology* 53 (3): 237–244.

77 Rio, S., Le Coq, L., Faur, C. et al. (2006b). Preparation of adsorbents from sewage sludge by steam activation for industrial emission treatment process. *Safety and Environmental Protection* 84 (B4): 258–264.

78 Chen, X., Jeyaseelan, S., and Graham, N. (2002). Physical and chemical properties study of the activated carbon made from sewage sludge. *Waste Management* 22 (7): 755–760.

79 Zhang, F.-S., Nriagu, J.O., and Itoh, H. (2005). Mercury removal from water using activated carbons derived from organic sewage sludge. *Water Research* 39 (2–3): 389–395.

2

Emerging Issues and Their Solutions Related to the Use of Sewage Sludge in Waste Treatment

Tanushree Paul[1] *and Arindam Sinharoy*[2,*]

[1] *Centre for the Environment, Indian Institute of Technology Guwahati, Guwahati, Assam, India*
[2] *Department of Microbiology, National University of Ireland Galway, Galway, Ireland*
* *Corresponding author*

2.1 Introduction

The alarming increase in the rate of sewage sludge generation worldwide is a matter of concern from an environmental point of view. Countries like China and India are among the largest producers of sewage sludge. Additionally, it has also been seen that in countries in the European Union, sewage sludge generation increased by 13 million tons by 2020 [1]. Sewage sludge was previously used for landfill application for absorbing the nutrients present in the sludge by the soil. But with the increasing population, the amount of land available for land-filling applications is decreasing. However, apart from the nutrient compounds present, sewage sludge also consists of certain micro-pollutants, hormones, and heavy metals, which certainly have detrimental effects on the environment. Hence, sewage sludge is termed hazardous owing to its composition consisting of heavy metals and pathogens. Furthermore, disposal of sewage sludge vectors sometimes produces an unpleasant odor, and generally produces toxic gases [2]. Furthermore, this practice might cause environmental problems by increasing groundwater contamination. Hence, there is a growing concern amongst environmentalists regarding the potential toxicity associated with untreated sludge generation. Hence, various techniques/technologies are being implemented for the reduction of potential threats caused by the extensive generation of sewage sludge. Apart from conventional techniques followed to date, researchers are focussed more on a sustainable approach, by converting the carbon content present in the waste which would significantly be converted into a suitable form of energy such as methane (or biogas) production using anaerobic digestion. Also, the recovered minerals and nitrogen from the digestate can also be used for fertilizers in this

Application of Sewage Sludge in Industrial Wastewater Treatment, First Edition. Edited by Maulin P. Shah.

process. However, the presence of different contaminants such as heavy metals, emerging pollutants and pathogenic microrganisms are a cause for concern for their safe processing and disposal. Hence, more research towards developing additional treatment methods is required for sewage sludge processing.

Different methods have been reported in the literature for sludge treatment which can help in removing these pollutants of concern. For example, Liew et al. [2] reported a detailed analysis of the stabilization of sewage sludge loaded with heavy metals by different thermal treatments. These thermal methods included incineration, gasification, pyrolysis, hydrothermal carbonization, and supercritical water gasification. In another study by Dubey et al. [3] more general treatment methods for sludge pretreatment were discussed which include sludge composting, anaerobic sludge digestion, ultrasonication, thermal hydrolysis, and ozonation. It also showed the transformation of other emerging contaminants and their conversion mechanism in detail. Also, vermicomposting and black soldier fly larvae treatment are some of the other biological treatments used for the treatment of sewage sludge.

2.2 Sewage Characteristics

Sewage sludge is the byproduct that is obtained when municipal wastewater is treated. Upon treatment, two main parts are obtained namely the aqueous part and the non-aqueous part which is termed sludge. During a typical wastewater treatment process, two types of sludge are generated known as primary and secondary sludge [2]. The composition of the primary and secondary sludge is shown in Table 2.1.

A typical flowchart of a municipal wastewater treatment plant is shown in Figure 2.1. Wastewater which is collected from different sources *viz* residential areas, industries, and restaurants is collected and fed into the initial screening tank for removing of unwanted materials [4]. This step is very important in order to reduce the damage which can be caused by these materials to the biological treatment system. The screened water is then allowed to flow to the grit chamber followed by the primary clarifier. The treatment process taking place in the primary clarifier is equally relevant as it provides a sufficient amount of hydraulic retention time for the settlement of the suspended solids. Various coagulants and flocculants are added in this step to further enhance the treatment efficiency in this process. The addition of coagulants in this step helps to neutralize the negatively charged colloidal particles, thus further facilitating the sedimentation process. Additionally, the flocculants also enhance this process by forming bigger flocs. These huge flocs easily settle at the bottom of the tank which is often termed the sludge cake. Furthermore, the top layer of the water in this tank will consist of insoluble light

Table 2.1 Chemical composition and properties of primary and secondary sludge [4].

Parameters	Primary Sludge	Secondary sludge
Total dry solids (TS, %)	1–6	0.4–1.2
Volatile solids (% of TS)	60–85	60–85
Grease and fats (% of TS)	5–8	5–12
Protein (% of TS)	20–30	32–41
Nitrogen (N, % of TS)	1.5–4	2.4–5
Phosphorus (P_2O_5, % of TS)	0.8–2.8	2.8–11
Potash (K_2O% of TS)	0–1	0.5–0.7
Cellulose (% of TS)	8–15	N/A
Iron (% of TS)	2–4	N/A
Silica (SiO_2, % of TS)	15–20	N/A
pH	5–8	6.5–8
Alkalinity (mg L^{-1} as $CaCO_3$)	500–1500	580–1100
Organic acids (mg L^{-1} as HAc)	200–2000	1100–1700
Energy content (kJ kg^{-1} VSS)	23000–29000	19 000–23 000

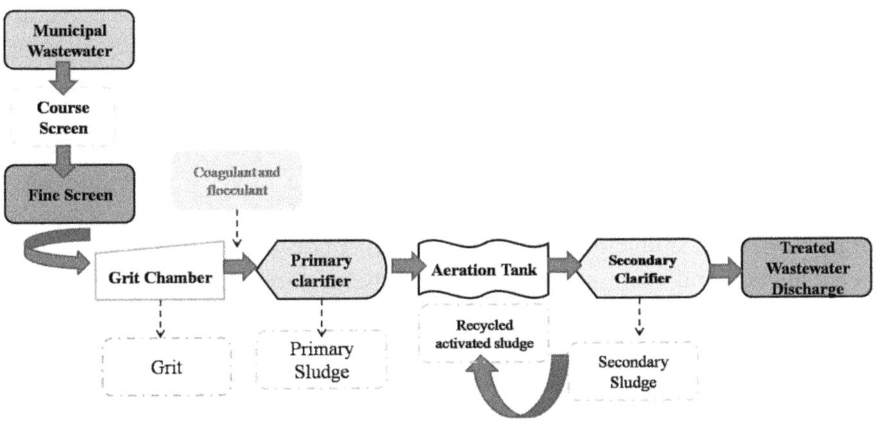

Figure 2.1 Flowchart showing different treatment steps involved in the treatment of municipal wastewater.

particles such as oils which could be removed by oil skimmers situated above the treatment tank and would be added to the sludge cake. This is now termed as the primary sludge. The output of this tank would be now allowed to enter the aeration tank which is the main process of the wastewater treatment plant. In this treatment process, biological treatment occurs with the presence of microbes which constantly take up the organics present in the wastewater thus reducing the nutrient (e.g. COD, nitrogen, etc.) content of the wastewater. The microorganism produced in this treatment tank is known as solids or sludge. In this case, the volume of mixed liquor suspended solids (MLSS) is calculated in order to identify the exact amount of secondary sludge to be removed from the tank.

The composition of the sludge is the main determining criteria for selecting the process to be used for its treatment. In this case the presence of heavy metals, emerging pollutants, and pathogenic microbes in the sludge should be estimated along with other physical and chemical parameters. A raw sewage sludge consists of 94% wt moisture [4]. Additionally, it is known that the largest fraction of the sewage sludge consists of proteins and polysaccharides [5]. The secondary sludge was also found to have a substantial amount of extracellular polymeric substances (EPS). The presence of this polymeric substance is equally important as its forms the backbone of the flocs and helps them to stick together, helps to absorb water, and enhances the proper functioning of the microbial population.

2.3 Emerging Issues in Sewage Sludge Treatment

The different emerging issues of sewage sludge processing are associated with the world wide industrial developmental problems and unchecked release of pollutants into the environment [6]. There are many pollutants which are part of the wastewater and concentrate in the sewage sludge, among these pollutants heavy metals and xenobiotic compounds have significant impact on the properties of the sludge and can potentially lead to serious environmental pollution [7]. The heavy metals such as cadmium, zinc, chromium, copper, lead, nickel, iron, etc. are commonly found in the sewage sludge [8]. The sources of heavy metals in sewage are from household, municipal, or industrial effluents [9]. The detection of these heavy metals and their quantification can be performed using standardized instrumental techniques such as energy dispersive x-ray spectroscope (EDX), x-ray photoelectron spectroscope (XPS), x-ray fluorescence (XRF) spectrometer, inductively coupled plasma – optical emission spectrometer (ICP-OES), and atomic absorption spectrometer (AAS) analyses [8]. In the case of these heavy metals existing in a crystalline form x-ray powder diffraction (XRD) technique can be used to detect them. These heavy metals have a toxic impact including an irreversible inhibitory effect on the microbial growth [10]. Moreover, when these heavy metals are present in the sludge, the processing of the sludge through anaerobic digestion is hampered due to their inhibitory effect on both methanogens and acetogenic microorganisms [11]. Some of the heavy metals are required in minute quantities for the microbial growth, however, they show serious inhibitory effects even at 10 mg L^{-1} concentration [12].

These metals can also be present in their nanoparticle form in the sewage sludge. The different nanoparticles are used in various medicines, personal care products, paints, and other formulations which become part of the sewage after their use and end up in the wastewater treatment systems within the sewage sludge [12, 13]. However, due to their higher activity they are sometimes even more toxic than the parent metals [14]. Once, they are released into the environment, they tend to persist in the environment and can potentially accumulate in the different biotic components of the ecosystem. These inorganic nanoparticles are carcinogenic, mutagenic, and tetargenic properties which makes them serious threat to living beings [15].

The xenobiotic compounds include medicine, cosmetics, biocides, and other personal care products. The main cause for release of these pollutants are widespread use of these compounds in human health care, from where they are released into the environment [16]. Among the medicines antibiotic and biocidic compounds present severe damage to the microbiome in the wastewater treatment systems and encourage the growth and proliferation of more resistant microbes (known as superbugs). Most of these compounds are hydrophobic in nature and tend to accumulate in the solid fraction i.e. the sludge [17]. This hydrophobic property along with their inhibitory nature make them difficult to degrade by microorganisms present in the biological treatment unit. This is the reason that these recalcitrant pollutants are known as persistent pollutants [18]. The detection of these pollutants in the sewage samples is really challenging due to their presence in minute quantities and the difficulty involved in their extraction. Different extraction methods including solid phase extraction, liquid liquid extraction, sonication aided extraction, etc. had been utilized for detecting xenobiotic and PCPs in sewage samples. The instrumentation used for their detection are mainly chromatographic techniques such as gas chromatograph (GC) and liquid chromatograph (LC) in combination with mass spectrometer (MS) [19].

2.4 Sludge Treatment Process

2.4.1 Pre-treatment

The entire sludge treatment process can be divided into pre-treatment, treatment, and disposal. The pre-treatment stage mainly emphasizes the removal of the moisture prior to the thermal treatment process such as incineration, pyrolysis, gasification, etc. Having said that, the presence of high moisture content can eventually lead to incomplete combustion, and hence it is essential to be removed [20]. High moisture from sludge can only be removed after mechanical dewatering of the sludge which enhances the moisture removal process. The disintegration of sludge particles is primarily targeted to enhance the overall degradability of sludge, making subsequent biological processes more efficient and effective [21]. It can be accomplished by other methods such as chemical, mechanical, enzymatic, and microwave treatment. The thermal treatment above 100 °C is considered to be the most efficient pre-treatment process in many cases. However, it continues to impart certain drawbacks due to high energy consumption [22]. The chemical pre-treatment is usually carried out in the presence of acids and alkalis which, although effective, may generate toxic byproducts that can harm the environment [23].

2.4.2 Thermal Treatment Process

The second stage of sludge stabilization is the thermal treatment stage which is performed after the pre-treatment process. It is mainly carried out to reduce the sludge volume, stabilize sludge, nutrient recovery, and subsequent generation of electricity. The main thermal techniques used in this stage include incineration, pyrolysis, gasification, supercritical water gasification, and hydrothermal carbonization. The thermal treatment process undoubtedly is more effective for the reduction of sludge volume. In the following section these thermal methods are discussed in detail along with their respective advantages and drawbacks (Figure 2.2].

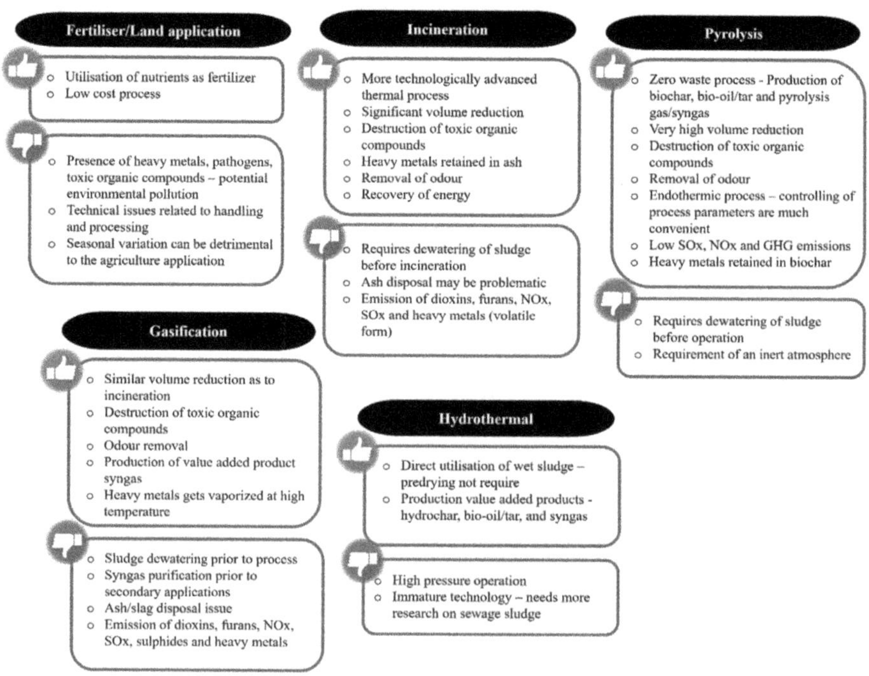

Figure 2.2 Comparison of different thermal treatment processes with conventional methods in terms of their advantages and disadvantages for sewage sludge disposal [Information *source*: 8].

2.4.2.1 Incineration

Incineration is the most widely used method for sludge treatment in European countries [24], and substantial research is now carried out in this area in order to improve the viability and efficiency of this process. In simple terms, sludge incineration is a basic combustion process that occurs in the presence of oxygen for the destruction of waste at an elevated temperature. This complex process not only generates carbon dioxide and water but other byproducts as well, which is dependent on several factors including incineration time, incineration temperature as well as the sludge composition itself. Incineration time and temperature are critical parameters in this case in order to facilitate complete combustion of sludge. Incomplete combustion might lead to the generation of certain unwanted toxic compounds such as CO. Insufficient oxygen supply might lead to partial oxidation and pyrolysis of the waste [25]. Metal oxides can also be formed if complete combustion doesn't take place in such a case. The incineration process consists of three basic steps, drying, devolatilization, and combustion. During the initial stage of this process, the pre-dried sludge is transferred to the incinerator and the temperature raised slowly to 100 °C as most of the heat is taken up by water as latent heat of vaporization [26]. After this, pyrolysis takes place followed by the last stage which is sludge char combustion leading to the generation of ash. From an industrial point of view, there are certain technologies available in this regard, some of which includes multiple hearth furnaces (MHF) and fluidized bed combustor (FBC).

Incineration is one of the most popular processes for sludge treatment as it significantly reduces the sludge volume to a small quantity which really helps in its easy disposal. The use of high temperature in this process also assures the destruction of harmful microorganisms thus reducing the probability of biohazards. If compared with landfilling process, it requires smaller space for ash disposal and also reduces the problem of spreading unpleasant odor. But, the pollutants which are produced during this process are certainly a concern to the environment. Release of NO_X, SO_X, dioxins and furans from the incineration process have been reported previously. In addition, the heavy metals present in the sludge get vaporized during incineration process, and once the system cooled down to environmental temperature, this vaporized metal contents gets cooled down to form ash particles.

Kozinki et al. [27] reported that about 78–98% of Cd, Cr, Cu, Ni, Pb, and Zn are retained in the ash itself, thus increasing the concern about the heavy metals as they can again leach out to the environment during disposal of the remains from the incineration process. Chen and Yan [28] examined the effect of the incineration temperature (500–1100 °C), time (0–60 min) and moisture content (0–75%) on heavy metal (Cd, Pb, Cr, Cu, Zn, Mn, and Ni) removal from sewage sludge collected from a sewage treatment plant located in Guangzhou province (China).

The results showed that the heavy metal removal behavior varied depending upon the metal species. For example, the Cd content in the solid phase decreased when the temperature raised from 500 to 1000 °C, but the Cu and Mn remained fixed. The maximum removal of Pb, Zn, Ni, and Cd as a solid fraction of incineration product was at temperatures of 900–1000 °C. The increased temperature showed a positive impact on the removal of these metals and although moisture content was important its effect was insignificant on the metal speciation.

2.4.2.2 Gasification

Another widely used technique for sludge treatment is gasification in which high-value synthetic gas production can be achieved. The gasification process is similar to incineration but it needs a high temperature for operation that is around 800–1100 °C and the process takes place in an oxygen deficit environment. This thermochemical conversion process takes place in different stages and it generates heat as the process is exothermic. The first stage includes the drying step (<200 °C) followed by the second step which focuses on devolatilization where the char and tar are formed. The third step includes the gasification of tar and char thus forming lighter gas particles. This process is also affected by various parameters *viz* temperature, gasifying agent residence time and use of catalyst. There are different gasifying agents like air, oxygen, steam, carbon dioxide, and other gas mixtures. Gasification in the presence of air is reported to produce the lowest LHV, whereas in the presence of oxygen and steam a comparatively higher LHV is produced [29]. However, gasification occurring in the presence of steam produces more tar as compared to oxygen, showing oxygen as the more suitable gasifying agent yielding higher carbon conversion [30]. Since the use of oxygen as a gasifying agent is considered less cost-effective, most of the research in this direction is carried out in the presence of steam [31]. Additionally, owing to a greater availability of water particles, the conversion through the water–gas shift reaction enhances. Due to this steam is the most extensively used gasifying agent in this regard.

Temperature, which is yet another crucial factor affecting the overall gas yield can be optimized for better productivity [32]. Another factor is the use of catalysts, which can directly affect product yield. Wang et al. [33] reported on a Ni/char catalyst which showed reduced tar production by approximately 987% while increasing its H_2 content. Dolomite, used as a catalyst can reduce tar production by 71% and similarly increase the H_2 content. The apparatus used during the gasification process is called a gasifier. There are different types of gasifiers, e.g. moving bed gasifier, fixed bed gasifier, and fluidized bed gasifier. The entire procedure for using a gasifier is reported in detail by Liew et al. [2] along with a critical analysis of each of the instruments. Of all of these, a fluidized gasifier is widely used due to the presence of a semi-suspended bed, being agitated by the gasifying agent.

However, one of the advantages of this process, as compared to incineration, is that it produces less SO_x and NO_x as it takes place in an oxygen defic environment This method can easily replace incineration as lesser pollutants are produced in this case. Nevertheless, the gasification process has its own drawback. Excessive tar formation during this process usually leads to clogging. However, this tar can be removed in the later part of the process or its production can be reduced within the gasifier itself. Similarly, as in the case of the incineration process, the heavy metals are vaporized at high temperatures forming metal compounds. The presence of a material bed in the gasifier, however, helps to absorb heavy metals thus reducing the heavy metal emission into the environment [34].

2.4.2.3 Pyrolysis

Another commonly used process for thermal treatment of sludge lies in the thermal decomposition of organic substances in the absence of oxygen, which is an endothermic process, requiring higher external energy for sustenance [35]. This process, however, necessitates lower moisture content in the sludge to proceed with the protocol, and due to this it is necessary to ensure better sludge drying and dewatering. It produces three main products namely biogas, biochar, and bio-oil and takes place in four main steps *viz* drying, devolatilization, secondary decomposition, and a final decomposition step. Temperature plays a critical role in this case for product formation. At a higher temperature, the amount of solid product decreases. At a lower temperature, below 650 °C, the liquid and gaseous component increases with an increase in temperature. Temperatures above this yield produce more of the gaseous products thus reducing the liquid product i.e. bio-oil content. Pyrolysis is primarily carried out to produce bio-oil which has the highest yield in this process, but the other byproducts such as biochar and aqueous phase (nutrient containing phase) can also be obtained through this process. Bio-oil, the main product, is a mixture of complex organic compounds, but is unstable due to the presence of high oxygen content, can be blended with conventional fossil fuel for use. The byproducts have their own uses which can further enhance the commercial value of the treatment process. For example, despite having less surface area than the commercially available adsorbents, biochar can be still used for adsorption. As far as heavy metal is concerned, almost 70% of heavy metals accumulate in the solid biochar [36]. Additionally, this method is reported to stabilize heavy metals at different temperatures [37].

Moško et ak. [38] studied the removal of recalcitrant pollutants such as PCBs, PAHs, pharmaceutical, endocrine disrupting, and hormonal compounds from anaerobically treated sewage sludge by pyrolysis at temperatures in the range 400–800 °C. The authors observed that only the pharmaceutical compounds could be removed at 400–600 °C pyrolysis temperature, whereas in order to remove the other compounds the reactor needed to be operated at a temperature higher than 600 °C. The biochar or the solid remains produced at such elevated temperatures are free from these pollutants, indicating the suitability of these materials to be disposed of in the environment without any threat of further pollution. Shao et al. [39] examined the distribution of three main heavy metals (Cu, Cr and Zn) in sludge from sewage treatment plants, paper mills, and slaughterhouses during the pyrolysis process. The authors found that the heavy metals moved towards more stable fraction i.e. the solid fraction of pyrolysis product. Moreover, due to their reduced risk of escaping into the environment, this method was found to be ideal for sewage sludge detoxification.

2.4.2.4 Supercritical Water Gasification and Hydrothermal Carbonization

Supercritical water gasification (SCWG) and hydrothermal carbonization (HC) are relatively new techniques being introduced for the thermal treatment of sludge. This technique is advantageous over other mentioned thermal techniques as it removes the necessity of the pre-drying step. It is basically a thermo-chemical process to convert organic materials into carbonaceous products, hydrochar in the presence of water. It usually takes place in a temperature range of 180–350 °C. Dehydration and decarboxylation reactions primarily take place in this process. The parameters which are studied for optimizing the process efficiency are residence time, temperature, moisture content, and catalyst [40]. The HC process occurs in sub-critical water conditions and removes the pre-drying step which is its prime advantage. As it is a comparatively new technique, more research is required in this area for the optimization of parameters for increasing hydrochar yield.

Liu et al. [41] reported that the maximum content of heavy metals such as Zn, Cu, Cd, Hg, Mn, and Pb is retained in the solid phase. As the temperature used in this process is usually kept below the boiling point of heavy metals, it is expected that the gaseous emission from this process would be free of heavy metals.

SCWG is another new hydrothermal process used for the conversion of feedstocks. In this method, water is used as the reactant in the form of supercritical water (SCW) which is achieved at 22.1 MPa and 374 °C. Owing to its low viscosity and highly diffusive nature it enhances the miscibility of the intermediate products formed during this process [42]. Like another thermochemical conversion process, this process also takes place in different steps which are thermal depolymerization, feedstock monomer degradation recombination of reactive fragments via polymerization, condensation and cyclization, followed by the last step which involves free radical reactions to form hydrocarbon gases [2]. In this type of process, temperature plays a critical role. It is reported that with an increase in the temperature H_2 gas production increases and as it dominates at 600 °C, it could be attributed to the fact that water–gas shift reaction is predominant in such cases due to the presence of steam formation. It also supports free radical reaction facilitating ring cleavage and forming low molecular gases. A substantial amount of research has been carried out in this regard following the study and effect of product distribution of sewage sludge, the effect of oxidation coefficient, and the effect of alkali and nutrient recovery [2, 43, 44]. This process is advantageous as it can start with wet sludge without pretreatment. The use of high pressure and temperature generates free radicals, and due to its high solvating power it can enhance the decomposition of organic compounds, further reducing the production of coal and tar. Although elimination of the intitial drying step reduces the operational cost in this method, however, the use of high-pressure equipment increases the overall capital cost. Heavy metals can be easily stabilized by this method as most of the heavy metals can be converted into stable forms of oxidizable reducible residue [45].

However, the bio-oil produced from hydrothermal treatment of sewage sludge contains very high amounts of heavy metals. For instance, Yuan et al. [46] reported 20–30 mg kg^{-1} of Cu and Cr, 10–20 mg kg^{-1} Zn, ~ 5 mg kg^{-1} Pb and Ni, and < 1 mg kg^{-1} Cd in liquefaction oil from sewage sludge. Among these metals, the Zn content in the bio-oil produced from sewage sludge at 360 °C was very high and can be detrimental to the environment if used in any application.

Hence, we can conclude that in all the thermal treatment methods discussed above, the dotoxification efficacy of sewage sludge using these methods was very high. The xenobiotics, micropollutants, nanoparticles, and heavy metals were either volatilized or retained in the solid fraction and there is very little chance of their being released into the environment. The micropollutants or the priority pollutants are completely degraded to their elements and mostly no trace of them could be found in the solid residue. In the case of heavy metals, almost > 70 wt% of them are reported to be retained in the end solid product. The heavy metals in all cases are easily converted into a stable form of oxidizable residue, and also it produces solid char that can further absorb the heavy metals. Further research is being carried out for the stabilization of heavy metals from sludge, however, it also depends on the initial concentration of heavy metal present prior to the treatment. In addition, the drug resistant microbes or superbugs could be removed completely along with the organic compounds present in sewage sludge by these thermal treatment methods. Table 2.2 depicts a summary of different thermal techniques and final products used.

2.5 Sustainable Future Prospective for Thermal Treatment of Stabilization of Sewage Sludge

Sewage sludge is a rich organic waste consisting of nutrients, recalcitrant compounds, and heavy metals. Each of these constituents need different suitable unit operations for its recovery and stabilization. It is essential to extract the valuable compounds of the sludge prior to landfilling or discarding. This would further

Table 2.2 Summary of different treatment methods and their performance criteria.

Treatment method	Condition	Moisture content	Other reactant	Main product	Secondary product	Carbon recovery in main product	Reference
Incineration	Oxygen excess; 800–1100 °C	<50 wt%	Air	Nil	Flue gas, ash	C recovery not the objective	[47]
Gasification	Oxygen limited; 650–1100 °C	<15 wt%	Air/steam/O_2	Syngas	Char, tar, water	C recovered in syngas	[48]
Pyrolysis	Oxygen deficit; 350–850 °C	<10 wt%	Inert gas	Bio-oil	Biogas, char	C recovery in bio-oil	[49]
HTC	140–350 °C; 2–10 MPa; presence of water	>80 wt%	Nil	Hydrochar	Process water, flue gas	C recovery in hydrochar	[50]
SCWG	Supercritical water state; > 374 °C; > 22.1 MPa	>80 wt%	Nil	Syngas	High TOC liquid, char	C and H recovery in syngas	[51]

reduce the harmful effects of such toxic compounds from the sludge. Sustainable state-of-the-art technologies are now a new research area in every aspect of this area. Additionally, the implementation of sustainable technologies in several industries further creates new job opportunities which also helps the development of developing nations. The heavy metals present in the sludge can be stabilized following thermal treatment techniques, this not only helps to create a sustainable approach towards nutrient recovery, but also helps in the reduction of harmful effects caused by the toxic elements present in the sludge which would contaminate the environment if sludge landfilling is done without treatment.

Among all the thermal treatment technologies used, while incineration can reduce the excellent volume reduction of sludge, other technologies such as pyrolysis and SCWG can also produce biofuel, thus efficiently recovering the carbon content of the sludge. Carbon content can also be recovered in the form of solid char which can be further used as an adsorbent for heavy metal absorption. All these technologies are advantageous as they require very short duration of time, however, the operation costs rise as high-tech equipment is required. Hence, ongoing research is focused on reducing the operational cost of these processes thus achieving a sustainable approach for overall nutrient recovery and sludge stabilization. New technology and process design can be implemented by using novel reactors and gasifiers for the reduction of tar. The use of a co-substrate, in this case, can further increase the process efficiency. Comparatively new technologies such as SCWG and HC require more optimization of process parameters to achieve excellent results. Furthermore, new treatment technologies can be developed for the treatment of sewage sludge, extracting the carbon content of the sludge along with value-added chemicals establishing a biorefinery approach. Recent literature suggests that some sludge can also be used for biochemical conversion and thus support the wastewater treatment process towards establishing a sustainable approach.

References

1 European Commission (2009). *Environmental, economic and social impacts of the use of sewage sludge on land*. 1–20.

2 Liew, C.S., Kiatkittipong, W., Lim, J.W. et al. (2021). Stabilization of heavy metals loaded sewage sludge: reviewing conventional to state-of-the-art thermal treatments in achieving energy sustainability. *Chemosphere* 277: 130310.

3 Dubey, M., Mohapatra, S., Tyagi, V.K., Suthar, S., and Kazmi, A.A. (2021). Occurrence, fate, and persistence of emerging micropollutants in sewage sludge treatment. *Environmental Pollution* 273: 116515.

4 Metcalf, L., Eddy, H.P., and Tchobanoglous, G. (2013). *Wastewater Engineering: Treatment and Resource Recovery*, 5e. Metcalf & Eddy, Inc.

5 Mikkelsen, L.H. and Keiding, K. (2002). Physico-chemical characteristics of full scale sewage sludges with implications to dewatering. *Water Research* 36: 2451e2462. doi: 10.1016/S0043-1354(01)00477-8.
6 Tarpani, R.R.Z., Alfonsin, C., Hospido, A., and Azapagic, A. (2020). Life cycle environmental impacts of sewage sludge treatment methods for resource recovery considering ecotoxicity of heavy metals and pharmaceutical and personal care products. *The Journal of Environmental Management* 260: 109643.
7 Raheem, A., Sikarwar, V.S., He, J. et al. (2018). Opportunities and challenges in sustainable treatment and resource reuse of sewage sludge: a review. *The Chemical Engineering Journal* 337: 616–641.
8 Udayanga, W.C., Veksha, A., Giannis, A. et al. (2018). Fate and distribution of heavy metals during thermal processing of sewage sludge. *Fuel* 226: 721–744.
9 Geng, H., Xu, Y., Zheng, L. et al. (2020). An overview of removing heavy metals from sewage sludge: achievements and perspectives. *Environmental Pollution* 266 (2): 115375.
10 Sinharoy, A., Pakshirajan, K., and Lens, P.N.L. (2020). Biological sulfate reduction using gaseous substrates to treat acid mine drainage. *Current Pollution Reports* 6: 328–344.
11 Smith, S.R. (2009). A critical review of the bioavailability and impacts of heavy metals in municipal solid waste composts compared to sewage sludge. *Environment International* 35 (1): 142–156.
12 Luo, J., Zhang, Q., Zhao, J. et al. (2020). Potential influences of exogenous pollutants occurred in waste activated sludge on anaerobic digestion: a review. *The Journal Hazardous Materials* 383: 121176.
13 Walser, T., Limbach, L.K., Brogioli, R. et al. (2012). Persistence of engineered nanoparticles in a municipal solid-waste incineration plant. *Nature Nanotechnology* 7 (8): 520–524.
14 Sinharoy, A. and Lens, P.N.L. (2020). Biological removal of selenate and selenite from wastewater: options for selenium recovery as nanoparticles. *Current Pollution Reports* 6: 230–249.
15 Wang, D. and Chen, Y. (2016). Critical review of the influences of nanoparticles on biological wastewater treatment and sludge digestion. *Critical Reviews in Biotechnology* 36 (5): 816–828.
16 Biel-Maeso, M., Corada-Fernández, C., and Lara-Martín, P.A. (2019). Removal of personal care products (PCPs) in wastewater and sludge treatment and their occurrence in receiving soils. *Water Research* 150: 129–139.
17 Ebele, A.J., Abdallah, M.A.E., and Harrad, S. (2017). Pharmaceuticals and personal care products (PPCPs) in the freshwater aquatic environment. *Emerging Contaminants* 3 (1): 1–16.
18 Byrns, G. (2001). The fate of xenobiotic organic compounds in wastewater treatment plants. *Water Research* 35 (10): 2523–2533.

19 Štefanac, T., Grgas, D., and Landeka Dragičević, T. (2021). Xenobiotics—division and methods of detection: a review. *Journal of Xenobiotics* 11 (4): 130–141.

20 Oladejo, J., Shi, K., Luo, X., Yang, G., Wu, T. (2019). A review of sludge-to-energy recovery methods. *Energies* 12 (1–38): doi: 10.3390/en12010060.

21 Ruffino, B., Campo, G., Genon, G. et al. (2015). Improvement of anaerobic digestion of sewage sludge in a wastewater treatment plant by means of mechanical and thermal pre-treatments: performance, energy and economical assessment. *Bioresource Technology* 175: 298–308. doi: 10.1016/j.biortech.2014.10.071.

22 Climent, M., Ferrer, I., Del Mar Baeza, M. et al. (2007). Effects of thermal and mechanical pretreatments of secondary sludge on biogas production under thermophilic conditions. *The Chemical Engineering Journal* 133: 335–342. doi: 10.1016/j.cej.2007.02.020.

23 Tulun, Ş. and Bilgin, M. (2019). Enhancement of anaerobic digestion of waste activated sludge by chemical pretreatment. *Fuel* 254: 115671. doi: 10.1016/j.fuel.2019.115671.

24 Kor-Bicakci, G. and Eskicioglu, C. (2019). Recent developments on thermal municipal sludge pretreatment technologies for enhanced anaerobic digestion. *Renewable and Sustainable Energy Reviews* 110: 423–443. doi: 10.1016/j.rser.2019.05.002.

25 Vallero, D. (2014). *Air pollutant emissions*. 10.1016/b978-0-12-401733-7.00029-3.

26 Klarschlamm, A.T.V. (1996). *ATV-Handbuch*, 4e. Berlin: Ernst and Sohn. Vesilind, P.A. (1994). The role of water in sludge dewatering. *Water Environmental Research* 66: 4–11. doi: 10.2175/wer.66.1.2.

27 Kozinki, J., Rink, K., and Lighty, T. (1995). Combustion of sludge waste in FBC. distribution of metals and particle sizes. In: *13th International Conference focussed on fluidized Bed Combustion* Orlando, FL: 305–315.

28 Chen, T. and Yan, B. (2012). Fixation and partitioning of heavy metals in slag after incineration of sewage sludge. *Waste Management* 32 (5): 957–964.

29 Nipattummakul, N., Ahmed, I.I., Kerdsuwan, S., and Gupta, A.K. (2010). Hydrogen and syngas production from sewage sludge via steam gasification. *International Journal of Hydrogen Energy* 35: 11738–11745. doi: 10.1016/j.ijhydene.2010.08.032.

30 Gil, J., Corella, J., Aznar, M.P., and Caballero, M.A. (1999). Biomass gasification in atmospheric and bubbling fluidized bed: effect of the type of gasifying agent on the product distribution. *Biomass Bioenergy* 17: 389–403. doi: 10.1016/S0961-9534(99)00055-0.

31 de Sales, C.A.V.B., Maya, D.M.Y., Lora, E.E.S. et al. (2017). Experimental study on biomass (eucalyptus spp.) gasification in a two-stage downdraft reactor by using mixtures of air, saturated steam and oxygen as gasifying agents. *Energy Conversion and Management* 145: 314–323. doi: 10.1016/j.enconman.2017.04.101.

32 Lee, U., Dong, J., and Chung, J.N. (2018). Experimental investigation of sewage sludge solid waste conversion to syngas using high temperature steam gasification. *Energy Conversion and Management* 158: 430–436. doi: 10.1016/j.enconman.2017.12.081.

33 Wang, D., Yuan, W., and Ji, W. (2011). Char and char-supported nickel catalysts for secondary syngas cleanup and conditioning. *Applied Energy* 88: 1656–1663. doi: 10.1016/j.apenergy.2010.11.041.

34 Nowak, B., Aschenbrenner, P., and Winter, F. (2013). Heavy metal removal from sewage sludge ash and municipal solid waste fly ash –a comparison. *Fuel Processing Technology* 105: Elsevier 195–201. doi: 10.1016/j.fuproc.2011.06.027.

35 Khiari, B., Marias, F., Zagrouba, F., and Vaxelaire, J. (2004). Analytical study of the pyrolysis process in a wastewater treatment pilot station. *Desalination* 167: 39-47.

36 Gao, N., Quan, C., Liu, B. et al. (2017). Continuous pyrolysis of sewage sludge in a screw-feeding reactor: products characterization and ecological risk assessment of heavy metals. *Energy Fuels* 31: 5063–5072. doi: 10.1021/acs.energyfuels.6b03112.

37 He, Y.D., Zhai, Y.B., Li, C.T. et al. (2010). The fate of Cu, Zn, Pb and Cd during the pyrolysis of sewage sludge at different temperatures. *Environmental Technology* 31: 567–574. doi: 10.1080/09593330903514466.

38 Moško, J., Pohořelý, M., Cajthaml, T. et al. (2021). Effect of pyrolysis temperature on removal of organic pollutants present in anaerobically stabilized sewage sludge. *Chemosphere* 265: 129082.

39 Shao, J., Yuan, X., Leng, L. et al. (2015). The comparison of the migration and transformation behavior of heavy metals during pyrolysis and liquefaction of municipal sewage sludge, paper mill sludge, and slaughterhouse sludge. *Bioresource Technology* 198: 16–22.

40 Paul, T., Sinharoy, A., Baskaran, D. et al. (2020). Bio-oil production from oleaginous microorganisms using hydrothermal liquefaction: a biorefinery approach. *Critical Reviews in Environmental Science and Technology* doi: 10.1080/10643389.2020.1820803.

41 Liu, T., Liu, Z., Zheng, Q. et al. (2018). Effect of hydrothermal carbonization on migration and environmental risk of heavy metals in sewage sludge during pyrolysis. *Bioresource Technology* 247: 282–290. doi: 10.1016/j.biortech.2017.09.090.

42 Zhang, L., Xu, C.C., and Champagne, P. (2010). Energy recovery from secondary pulp/paper-mill sludge and sewage sludge with supercritical water treatment. *Bioresource Technology* 101: 2713–2721. doi: 10.1016/j.biortech.2009.11.106.

43 Amrullah, A. and Matsumura, Y. (2018). Supercritical water gasification of sewage sludge in continuous reactor. *Bioresour. Technol* 249: 276–283. doi: 10.1016/j.biortech.2017.10.002.

44 Qian, L., Wang, S., Xu, D. et al. (2015). Treatment of sewage sludge in supercritical water and evaluation of the combined process of supercritical water gasification and oxidation. *Bioresource Technology* 176: 218–224. doi: 10.1016/j.biortech.2014.10.125.

45 Xu, Z., Zhu, W., Bao, J., and Chen, J., (2011). The fate of heavy metal during subcritical and supercritical water gasification of sewage sludge. *ISWREP 2011 – Proceedings. 2011 International. Symposium. Water Resource and Environmental Protection* 2: 1260–1263. doi: 10.1109/ISWREP.2011.5893247.

46 Yuan, X., Leng, L., Huang, H. et al. (2015). Speciation and environmental risk assessment of heavy metal in bio-oil from liquefaction/pyrolysis of sewage sludge. *Chemosphere* 120: 645–652.

47 Hao, Z., Yang, B., and Jahng, D. (2018). Combustion characteristics of biodried sewage sludge. *Waste Management* 72: 296–305. doi: 10.1016/j.wasman.2017.11.008.

48 Choi, Y.K., Ko, J.H., and Kim, J.S. (2018). Gasification of dried sewage sludge using an innovative three-stage gasifier: clean and H_2-rich gas production using condensers as the only secondary tar removal apparatus. *Fuel* 216: 810–817. doi: 10.1016/j.fuel.2017.12.068.

49 Tomasi Morgano, M., Leibold, H., Richter, F., Stapf, D., and Seifert, H. (2018). Screw pyrolysis technology for sewage sludge treatment. *Waste Management* 73: 487–495. doi: 10.1016/j.wasman.2017.05.049.

50 Danso-Boateng, E., Shama, G., Wheatley, A.D., Martin, S.J., and Holdich, R.G. (2015). Hydrothermal carbonisation of sewage sludge: effect of process conditions on product characteristics and methane production. *Bioresource Technology* 177: 318–327. doi: 10.1016/j.biortech.2014.11.096.

51 Hantoko, D., Antoni, Kanchanatip, E. et al., 2019. Assessment of sewage sludge gasification in supercritical water for H2-rich syngas production. *Process Safety and Environmental Protection* 131: 63–72. 10.1016/j.psep.2019.08.035.

3

A Detailed Overview of Anaerobic Digestion of Sewage Sludge with Different Process Intensification Strategies

Bharat Bhushan Negi[1] *and Arindam Sinharoy*[2,*]

[1] *Department of Chemical Engineering, Indian Institute of Technology Guwahati, Guwahati, Assam, India*
[2] *Department of Microbiology, National University of Ireland Galway, Galway, Ireland*
* *Corresponding author*

3.1 Introduction

The freshwater component of the world is around 2.5 %, with glaciers accounting for a large proportion of it. The ecology and climate change have been negatively impacted by urbanization, industrialization, and the expansion in the human population [1]. The environment has been harmed by the direct dumping of wastewater effluent from industrial and domestic sources. Organic and inorganic components, as well as heavy metals, can be found in municipal wastewater, anaerobically digested wastewater, industrial wastewater, and agricultural wastewater [2]. These wastewaters are treated using physicochemical and biological approaches in a multistage process. Biological treatment methods use biological agents (bacteria, fungi, and algae) to treat wastewater [3]. When using bacteria in the common aerobic wastewater treatment method such as the activated sludge process, the organic component of the wastewater is converted to biosolids or sludge (Figure 3.1). The commonly used disposal methods for sewage sludge are agricultural spreading, landfills, composting, anaerobic digestion, recycling as a construction material, and incineration [4, 5].

Among those processes anaerobic digestion is a promising alternative for the disposal of such organic waste. The use of anaerobic digestion to treat and valorize waste biomass is beneficial for not only disposing of the waste but also generating alternative sources of energy [6, 7]. Anaerobic digestion is also clean and renewable, and an alternative energy source to the rapidly depleting supplies of conventional fossil fuels. In the absence of oxygen, anaerobic digestion degrades organic materials. It is a natural process carried out by a wide range of microbial

Application of Sewage Sludge in Industrial Wastewater Treatment, First Edition. Edited by Maulin P. Shah.

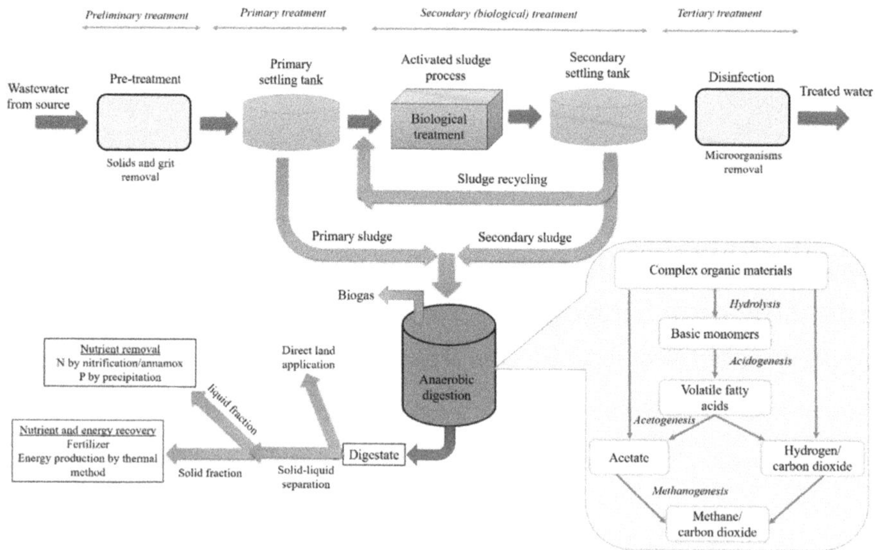

Figure 3.1 Overview of a conventional wastewater treatment plant along with waste sludge generation integrated with its anaerobic digestion (including processes involved and products formed).

communities found in various environmental settings mostly belonging to the anaerobic group of microbes. Anaerobic digestion creates biogas from the decomposition of organic waste [8], which contains CH_4 (60–70%), CO_2 (30–40%) and less than 1% nitrogen [9]. This technique has attracted a lot of interest in both scientific research and practice over the last decade. However, the efficiency of anaerobic digestion may be limited by a lack of waste diversity from a single source, which is insufficient for large-scale digesters and the disadvantages of using single substrates, such as improper carbon–nitrogen (C/N) ratios, low substrate pH, and high ammonia concentrations. As a result, co-digestion of mixed substrates for biogas production which leads to the correct balance of nutrients has attracted a lot of attention recently [10].

3.2 Anaerobic Digestion of Sewage Sludge

Anaerobic digestion is the biological process of the degradation of organic matter without oxygen. This process takes place with the help of anaerobes, i.e. acetogenic bacteria, methanogenic bacteria or a consortium of microorganisms [9]. The method of the anaerobic digestion of sludge involves the breakdown and destruction of the degradable compound into methane and carbon dioxide gas in the absence of oxygen [11], it removes around 50% of organics from wa activated sludge; therefore, it is considered to be the most beneficial technology for sludge reduction and waste stabilization [12]. Many microorganisms have been identified as responsible for this process [13], which plays a significant role in environmental sustainability by valorization of waste into energy [14]. The reduction process takes place in different biochemical reactions under anoxic conditions. Therefore, it is considered to be a complex process because sewage sludge consists of a mixture of components mainly in particulate forms, such as lipids, proteins, and carbohydrates, which have different biodegradability characteristics. Microorganisms cannot directly utilize complex organics; therefore, they are transformed into simple mono or oligo forms, which involves hydrolysis, acidogenesis, acetogenesis, and methanogenesis reactions (Figure 3.1).

Hydrolysis is the first step of anaerobic digestion, which involves the enzyme-mediated transformation of complex organics materials such as proteins, fats, lipids, and polysaccharides into the simplest form, which the microorganisms can directly take. It is a slow process carried out by a large group of facultative bacteria and clostridia, which can grow in anaerobic conditions [15]. These microorganisms secrete certain types of extracellular enzymes such as cellulase, protease, and lipase; different microorganisms secrete different enzymes, which breaks either sugar or protein; there are different enzymes for sugar, fats, and proteins. Those who break sugar are called saccharolytic, and those who break proteins are called proteolytic microorganisms; they catalyze a large number of organic compounds into simple sugars, which microorganisms can take into the cell as a source of energy and nutrition [16]. This is a rate-limiting step, which depends upon the nature of the particulate matter (availability of free accessible surface area) [11] and the overall structure of the organic compound. Hydrolysis can be enhanced by increasing the temperature of the digester [12].

The monomer produced from the hydrolysis of organic compounds is utilized in acidogenesis reactions. Where various facultative microbes act on these monomers and degrade them into short-chain fatty acids, acetate, butyrate, aldehyde, alcohol, and gases such as carbon dioxide and hydrogen. This fermentation process starts with simple glycolysis where glucose breaks down into the pyruvate and formation of NADH occurs. The formation of hydrogen is catalyzed by hydrogenase enzyme (metalloenzyme). The amount of hydrogen released at the end decreases with the increase in the amount of non-gaseous products formed. Acetate, hydrogen, and carbon dioxide can be directly utilized for the last step of anaerobic digestion by methanogens in the methanogenic reaction [17]. High partial pressure of hydrogen in this step can decrease the amount of reduced compounds [6]. Acetogens further utilize the short chain fatty acids (except acetate) and converts them into organic acids such as, butyric acid, acetic acid, propionic acids, and hydrogen and carbon dioxide gases, this process is known as acetogenesis. Two types of bacteria, homoacetogen and hydrogen-producing acetogens, are involved in this process which provides the substrate for methane production. At the end of digestion, methane and carbon dioxide formation occurs with the help of acetoclastic and hydrogenotrophic methanogens, which is known as methanogenesis. Acetoclastic methanogens convert the acetate into methane and carbon dioxide, whereas hydrogenotrophic methanogens form methane from dissolved hydrogen and carbon dioxide gas. This step is the most critical and slowest in biochemical reactions.

In addition to the production of biogas (or methane), the anaerobic digestion process produces digestate which is nutrient rich compound, which can be directly applied as organic fertilizer (Figure 3.1). The amount of digestate is very large as almost 90% of the substrate that is fed to the anaerobic digester is retained as digested. It can also be fractionized into solid and liquid digestate using a solid–liquid separation step, and all of this fraction can be applied directly to the land as fertilizer. The liquid fraction contains mostly N and P components of the digestate which can be separated using biological (e.g. nitrification or Anammox process for N removal) or chemical (e.g. P precipitation for its recovery). The solid fraction can be treated using thermal methods (e.g. pyrolysis, hydrothermal treatment, gasification, etc.) for energy recovery and reduction in its volume [18, 19, 20].

3.3 Factors Affecting Sludge Digestion

Several factors such as pH, temperature, carbon to nitrogen ratio, hydraulic retention time, and microbes used significantly affect the anaerobic digestion process and its efficiency. Anaerobic digestion is a slow process and different microorganisms are involved in the whole biochemical process. During treatment of sludge containing a high number of total ammonia nitrogen (TAN) produced from biodegradation of nitrogenous matter affects the pH [21] the growth rate of bacteria is affected by the pH change; therefore the pH level needs to be adjusted to ensure successful digestion. The value for pH in an anaerobic digester is primarily neutral; the optimum pH range for methanogenic microorganisms is 6.5–8.0, whereas the bacteria responsible for fermentation functions at an optimum pH range of 5.0–6.0 [9]. Inhibition of pH and digestion process occurs due to toxic substances such as ammonia, sulfide, and heavy metals in the waste stream of different origins [21], therefore a prior knowledge of waste components is essential to complete successful digestion.

Temperature is also a necessary factor in sludge digestion; a high rate of anaerobic digestion occurs with rapid hydrolysis. Anaerobic digestion occurs at different temperatures ranges, i.e. psychrophilic (< 20 °C), mesophilic (25–40 °C), and thermophilic (above 45 °C), increasing the temperature of the digester to thermophilic condition (55–70 °C) shows higher hydrolysis and faster reaction rate in comparison to the mesophilic state (37 °C) [12]. At a low temperature, below 20 °C, the process becomes challenging due to a change in the physicochemical nature and slow biochemical reaction of the sludge [22]. Acid forming bacteria primarily functions at the mesophilic temperature, whereas a high temperature is required for methanogens. Increasing the temperature gives rise to a more hydrolyzed soluble organic compound, which is more accessible for microorganisms. However, a high temperature leads to acidification and an increase in the fraction of free ammonia, which inhibits bacterial growth and affects the digestion process. The methanogenesis and process stability decreases in the thermophilic system, and it requires large investment, whereas, in the mesophilic system, process stability is better and affords low methane yield [23].

The anaerobic digestion process can be improved by lowering the hydraulic retention time (HRT) during the thermophilic condition [12]. HRT is a crucial factor for the slow growing microbes in the reactor such as *Methanosarcina*. The doubling time of methanogens are very slow (5–16 days) [16], therefore the HRT should be 10–15 days for the process.

All the organisms present in the system need essential elements for their growth and performance, nitrogen is amongst the most important nutrient element which is responsible for amino acid and protein synthesis. All of the feedstock contains a set carbon to nitrogen (C/N) ratio and other nutrients such as phosphorus and sulfur. Substrates having a low C/N causes insufficient utilization of the carbon source, and a high amount of ammonia which inhibits the microbial growth, therefore, ammonia inhibition during this process can be avoided by increasing the C/N ratio to an appropriate level [10]. Sometimes a high C/N ratio causes low nitrogen in the medium and fast degradation of nitrogen leading to poor biogas production [23]. Hungría et al. [8] reported the highest methane production during co-digestion at a C/N ratio of 4/5, the digestion system is sensitive to the C/N ratio, high C/N decreases the total ammonia nitrogen (TAN) due to low protein stabilization and is an advantageous for the system [10].

3.4 Bioreactors for Sludge Digestion

3.4.1 Different Anaerobic Bioreactor Configurations

Different bioreactor configurations, namely stirred tank reactor, anaerobic contract reactor, upflow anaerobic sludge blanket reactor, anaerobic baffled reactor, sequencing batch reactor, anaerobic filter, expanding bed reactor, fluidized-bed reactor, and anaerobic membrane bioreactor, have all been explored for anaerobic digestion. The stirred tank type configuration is the most common type of bioreactor, not only for anaerobic digestion but also for other types of bioprocesses. The content of the reactor is continuously mixed throughout the reaction period using mostly a mechanical agitator. Recently other strategies such as intermittent mixing using treated water recirculation or gas sparging was used to achieve mixing inside the reactor. An anaerobic contract reactor is a modification of a stirred tank type bioreactor with an added settling tank associated with the main digester where the solids can attach to the microbes to form flocks and settle easily. The settled biomass is often recycled back to the main reactor for biodegradation. This type of bioreactor is more suited to the treatment of high solid containing waste streams and can achieve high treatment efficiency.

An upflow anaerobic sludge blanket reactor is another commonly used bioreactor for anaerobic digestion of sewage sludge. The main feature of this reactor is the formation of sludge granules which is advantageous for biomass retention inside the bioreactor to achieve high conversion rates and biogas yield. Although designed for treating low COD wastewater as their residence time is quite low (few hours), they have also been adapted for treating high organic loading content wastewater. An anaerobic baffled reactor consists of a larger reactor divided into smaller chambers using baffles which make the wastewater travel in a zig-zag pattern. This type of flow regime inside the bioreactor makes the solid retention longer and results in biomass concentration in the bioreactor. The mechanism in a baffled reactor is a combination of the stirred tank and plug flow type reactors with very high treatment efficiency. The anaerobic sequencing batch reactor is operated under a different sequence of operations, namely, fill, react, settle, and decant. This type of setup reduces land use by operating a single reactor in four different conditions. The settle stage provides a very high degree of solid–liquid separation by allowing the biomass flocks to settle and this in turn provides a better quality effluent.

Anaerobic filters consist of a filter bed with different biosupport materials upon which the microbial biofilm grows. The wastewater is supplied either in an upflow or a downflow mode and the biofilm present inside the reactor degrades the organic components. This type of setup is not suited for sewage sludge treatment as there is high possibility of clogging the pores through which the wastewater percolates. This reactor is most suitable for wastewater containing low volumes of solid. In order to avoid filter bed clogging the back washing or reversing the flow (in the case of downflow reactor) can provide good results. Expanding bed and fluidized bed reactors have similar principles of operation, in which the influent wastewater stream fluidizes the bed supporting the biomass. The biomass can be grown as a biofilm onto support materials in both cases, however, for the expanding bed the biomass can also be in the form of granules. In order to achieve the desired fluidization the treated effluent can also be recycled back into the reactor along with influent flow. These two types of reactor are best for high COD or high organic containing wastewater and provide better treatment efficiency values.

3.4.2 Membrane Bioreactor for Sewage Sludge Digestion

Although membrane bioprocess technology is not new, in recent times it has gained a lot of traction from researcher and industry alike for treatment of different industrial wastewaters. An anaerobic membrane bioreactor has also been effectively used for digesting sewage sludge. For this purpose various membrane bioreactor configurations and different operational conditions have been tested. In the following use of different membrane bioreactors for sewage sludge digestion and biogas production will be discussed with appropriate examples.

Qin et al. [24] studied the co-digestion of sewage sludge along with food waste in an immersed membrane bioreactor system under mesophilic conditions (37 °C). The study found a very good treatment efficiency with high volatile fatty acid production in the bioreactor. The volatile fatty acid production and its composition was dependent upon the microbial community, which in turn correlated with the organic loading in the bioreactor. Under optimum conditions, microbes belonging to *Firmicutes* and *Actinobacteria* were dominant in the biofilm formed. In a similar study, a membrane bioreactor was used for co-digestion of food waste and sewage sludge under thermophilic conditions (55 °C). The reactor could be successfully operated at a very high solids loading (27 g L^{-1}). The co-digestion strategy obtained a very high biogas of 0.8 g L^{-1} of VSS which was 91–158% higher than the values obtained for monodigestion of either individual waste.

A study with a forward osmotic membrane bioreactor for anaerobic digestion of sewage sludge showed great potential for higher biogas yield and methane concentration [25]. The forward osmosis was used to dewater the sludge and helped to maintain high solid content in the reactor. The membrane reactor could remove 60% of the organics from the sewage sludge which was 9% higher than the control reactor without membrane. Song et al. [26] treated sewage sludge using an osmotic membrane bioreactor. The changes in the rheological property of the sludge due to anaerobic digestion in a membrane bioreactor was compared with a conventional sludge digester. The membrane bioreactor treated sludge showed better rheological property which according to authors can save up to 39% of energy consumption required for sludge mixing. Moreover, the solid content in the bioreactor was also higher at ~ 7.5% up from ~ 3.5% in a conventional anaerobic digester.

Wandera et al. [27] treated a biologically pretreated (under thermophilic conditions) sewage sludge using an anaerobic membrane bioreactor. This essentially made the system a two stage process, where the first thermophilic (70 °C) stage with CSTR was for hydrolysis and second mesophilic (37 °C) membrane bioreactor for methane generation. The reactor could achieve good amount of solid removal (46–63%) along with high methane production (1.3 L L^{-1} d^{-1} at 5 d HRT). After a prolonged reactor operation, the authors encountered membrane fouling which they could effectively remove using chemical treatment such as a combination of sodium hypochlorite and citric acid. In a similar study, thermal pretreated sewage sludge was anaerobically digested in a membrane bioreactor under mesophilic conditions [28]. Among the different thermal conditions ranging from 65 °C to 125 °C, 125 °C was optimum for hydrolysing the sludge effectively with a 156% increase in the soluble COD value. The methane content in the biogas produced in the membrane bioreactor increased from 47.9% for untreated sludge to 68.3% for sludge treated at 125 °C. The COD removal also nearly completed in different stages of bioreactor operation indicating good performance criteria for the membrane bioreactor.

3.4.3 Pilot-scale Anaerobic Digestion

Pilot-scale studies with a bioreactor is an important aspect for the commercialization of any technology. This is an intermediary step between lab-scale and full commercial-scale reactors in size and capacity. The knowledge gained during lab-scale studies are utilized in designing a pilot-scale system and again information obtained from this helps in the design of a full commercial-scale bioreactor system. Due to their smaller size compared with full-scale plants, monitoring key parameters are much easier, and based on the information obtained carrying out modification of the bioreactor is also easy. This way pilot-scale studies are an essential step towards full-scale plants. There are quite a few studies on pilot-scale sewage sludge digestion which provides great insight into the process design in this area of research.

For instance Lopez et al. [29] compared three different pilot-scale bioreactors (a thermophilic anaerobic, a mesophilic anaerobic, and a thermophilic aerobic reactor) for sewage sludge digestion. The aerobic reactor had a volume of 3.9 m^3, whereas both the anaerobic ones have a volume of 2.75 m^3. Both anaerobic reactors performed much better than the aerobic reactor in terms of organic load removal. The organic load removal values for anaerobic mesophilic, anaerobic thermophilic, and aerobic mesophilic reactors were 0.54–0.8, 0.69–1.53 and 0.24–0.39 kg VSS m^{-3} d^{-1}, respectively. Moreover, the anaerobic reactor produced biogas a value added product from sewage sludge with thermophilic reactor performing best with 1730–2790 L d^{-1} of biogas production rate.

There were many pilot-scale studies on co-digestion of sewage sludge with other organic wastes such as food waste, kitchen waste (fruit, vegetable, etc.), agro-industrial wastes, organic domestic waste, etc. This co-digestion strategy was found to be better than mono-digestion as it helps in balancing the nutrients required for optimum microbial activity. Furthermore, co-digestion helps with adjusting water content in the substrate, balancing pH, diluting the toxic compounds present in one of the wastes, and improving bioavailability of the substrates to the microorganisms. In a study by Zupančič et al. [30], the sewage sludge is co-digested with domestic organic wastes in two mesophilic digesters with a total volume of 2000 m^3 and at 20 d HRT. The addition of domestic organic waste improved the biogas production by 56% and biogas quality (methane content) by 80%. A further 10% increase in the VSS degradation was also observed in co-digestion over mono-digestion with sewage sludge.

Maragkaki et al. [31] studied co-digestion of sewage sludge with olive mill wastewater, cheese whey, and crude glycerol in a 220 L volume pilot-scale reactor. The results demonstrated that addition of these agro-industrial wastes to sewage sludge even at a low concentration of 5% (by volume) could cause a significant increase (220%, 350%, and 86% for olive mill, glycerol, and cheese whey, respectively) in biogas production. The maximum methane production of 185.7 L d^{-1} was with 5% crude glycerol. In another study, co-digestion of municipal waste with waste activated sludge was carried out in a pilot-scale reactor of 2 m^3 total volume [32]. The maximum methane production rate was 2.94 m^3 m^{-3} d^{-1} at an organic loading rate of 8.0 kg VSS m^{-3} d^{-1} and 15 D of HRT. The increase in the organic loading to the reactor adversely affected the bioreactor performance with poor solid removal, low methane production, and an increase in acidogenesis.

The combination of different pretreatment strategies with anaerobic digestion of sewage sludge has also been explored in some pilot-scale studies. The objective of such combined treatment is improved sludge digestion and an increase in product formation from sewage sludge. For example, in a study of hydrothermal treatment, anaerobic digestion and pyrolysis was sequentially applied to sewage sludge for complete utilization of sewage sludge [33]. In the first step, hydrothermal treatment at 180 °C for 30 minutes followed by filtration removed more than 60% of the water content in the sludge. The filtrate was subjected to anaerobic digestion under mesophilic conditions and the solid sludge from filtration process treated using pyrolysis 600 °C. From this process biogas and biochar were obtained as valuable products. In a similar study, hydrothermal or alkaline hydrothermal pretreatment was combined with anaerobic digestion [34]. The study found that an alkaline hydrothermal treatment using $Ca(OH)_2$ was better than just a hydrothermal treatment for sludge dewatering in the next step. The increase in hydrothermal treatment had a positive effect on the performance but reaction time did not significantly impact sludge digestion. The hydrothermally treated sewage was further dewatered and the filtrate was subjected to an anaerobic process for biogas generation. The authors suggested that the produced biogas could be utilized for steam generation which could be used in the thermal process making the proposed system a circular process with better utilization of waste.

3.4.4 Full-scale Anaerobic Digestion of Sewage Sludge

The utilization of the anaerobic digestion process for treating excess sludge generated in conventional wastewater treatment plants is quite common practice. In many places worldwide such installations can be found. Hence, there is no dearth of studies on full-scale anaerobic digesters for treating sewage sludge. Here a few of the important studies are discussed to provide readers with sufficient information about this aspect of sludge digestion.

Lloret et al. [35] explored the performance of a full-scale anaerobic digester (15 m^3, CSTR) over a long period (one and half years) with the key focus on removal of pathogenic organisms. Both mesophilic (35 °C) and thermophilic (55°C) conditions were examined to observe their effect on the digester performance. Both the reactors performed well in terms of solid removal and biogas production, however, the thermophilic one provided better stability even at high organic loading conditions and also at low retention times. The pathogenic organisms were not detected in the thermophilic digested sludge and heavy metals within the treated sludge were below detection level, indicating their suitability for safe disposal.

In a full-scale anaerobic digester a co-digestion approach with sewage sludge and organic fraction of the municipal solid waste enhanced the methane production with values reaching upto 90 % with additional feed [36]. The study found that the sludge from co-digestion was more active than the sludge from digester treating only sewage sludge, confirming the reason behind better methane production in a co-digestion plant. The methane production was greatly affected under substrate limiting conditions, hence, the methane yield varied in different cycles with different feeding conditions.

Modelling of a full-scale bioreactor is a rather less explored area of research in the field of sewage sludge digestion. A study from the UK tried to investigate this with the help of data obtained from 72 full-scale anaerobic sludge digesters collected over a period of seven years [37]. This work created a multi-level regression-based model that could predict the digester performance from measurable parameters such as temperature, hydraulic retention time, and solid concentration. The results of the study revealed that over-reliance on biogas yield to predict the digester performance is not the correct approach, rather focus should be placed on the correlation between the solids in feed, retention time, and temperature with biogas production. This comprehensive approach can help in more accurate prediction of performance.

Microbes play a critical role in anaerobic digestion and understanding their population dynamics inside a digester can provide great insight into the digester performance. Furthermore, the effect of different operating parameters such as organic load, solid content, pH, temperature, etc. can help in forming the microbiome in a digester, and hence, the role of all such parameters needs to be understood. This will not only help in improving the digester performance but will also help in understanding reactor failures that may occur and prevent them. There are a number of studies aiming to understand the microbial population inside an anaerobic digester treating sewage sludge under different operating conditions.

Lee et al. [38] examined the microbial community in 10 full-scale anaerobic digesters treating either food waste or sewage sludge in order to understand the effect of total ammonia nitrogen and sodium ions on them. Although, both these compounds showed significant effects on the microbial community formation in both types of digester, the high ammonia nitrogen concentration showed detrimental results on biogas production. For both the food waste and sewage sludge digesters, the methanogenic species were dominant (99.8% and 98.4%, respectively). Among the methanogenic archaea, *Methanoculleus* sp. was most prominent in the food waste digester, whereas, in the sewage sludge digester two species of *Methanolinea* and *Methanosaeta* sp. were the dominant species. Moreover, hydrogenotrophic methanogens were predominant in both systems with 96.9% and 79.4% of these specific strains present in food waste and sludge digester, respectively. The bacterial community was more diverse with species capable of organic compound degradation present in both systems.

In a similar study, the effect of process parameters on microbial communities in four large-scale anaerobic digesters treating waste activated sludge was examined [39]. In this study, the two digester operated at a slightly lower pH of 6.1–6.4 and showed higher bacterial diversity and more microbial community than the other two digester that are at pH 7.2–7.4. The dominant microbial population belongrf to the fatty acid or protein/amino acid-utilizers which could be important for the digester performance. The microbial population dynamics correlated well with one or more measurable process parameters such as pH, ammonia or potassium ions, lipid removal efficiency, etc. This findings will help in selective enrichment of a specific microbial community for improving digester performance.

An interesting finding by Wang et al. [40] showed that the microbial community inside an anaerobic digester was shaped by influent microbial species in sewage sludge that is being digested. In this study the microbial community in an anaerobic digester treating a mixture of primary sludge and waste activated sludge was analysed. The results showed the archaeal community is similar in both the feed sludge and the digested sludge, however their abundance varied greatly. The fungal species in the digested sludge was the same as those present in waste activated sludge but not present in the primary sludge, indicating the source of the fungal community inside the anaerobic digester. Moreover, there were many unknown microbial species present in the sludge among which many could potentially play an important role in the different processes in anaerobic digestion.

Kim et al. [41] analysed the microbial community in three full-scale anaerobic sewage sludge digesters with 164 000, 41 000 and 308 000 $m^3\ day^{-1}$ capacity. All the digesters were modelled as stirred tank type configurations and operated at mesophilic temperatures. The average solid removal efficiency was 50–55% and the methane content in the biogas produced was 60–65%. The hydrogenotrophic methanogens were dominant in all the plants with more than 94% of the methanogenic species belonging to this category. Moreover, the results indicated that the majority of the bacterial species were unknown and all of the archaeal species belong to methanogenic groups. These findings contradicted with the conventional knowledge that acetoclastic organisms are the dominant methanogens species in anaerobic digesters, hence, the authors suggested that the parameters deciding the microbial community dynamics should be further studied to determine the reasons behind their findings.

The anaerobic digestion is the only wastewater treatment technology that can potentially achieve net energy-neutrality, i.e. break-even the energy input though biogas production. In a work by Jenicek et al. [42] the data from a full-scale anaerobic digester is used to find the possibility of becoming energy self-sufficient in a wastewater treatment system by the energy generated in the form of biogas. The study found that this is possible but the key to achieving this feat is to optimize the energy utilization in such a plant and to reach a biogas yield of 600 L kg^{-1} of VSS fed to the digester. Just by fulfilling these two criteria a wastewater treatment plant could attain self-reliance in terms of energy consumption even if the plant is located in a modest climate zone.

3.5 Process Improvement Strategies

Although, the anaerobic digestion of sewage sludge is an effective method for its disposal in addition to being an alternative method for generating bioenergy and other value added products, application of this technology is rather hindered by to the complex nature of the sewage sludge. This difficult cellular structure combined with its complex composition poses great challenges to its biodegradation during anaerobic digestion, and hence requires strategies to improve its biodegradability. For this the most common strategy is pretreatment of sewage sludge using different physical, chemical, and biological methods or different combinations of them. These methods break down the complex structure of the sludge and makes it easy for microbes to act upon. This type of pretreatment not only improves process efficiency for biogas production and overall yield of product on substrate, but also reduces residence time inside the reactor by which the process cost can be further improved.

3.5.1 Sludge Pretreatment

3.5.1.1 Physical and Mechanical Treatment

Now we discuss the different pretreatment technologies that are available for improving anaerobic digestion of sewage sludge. The first method is the physical or mechanical treatment methods which are grinding (ball milling), sonication, microwave irradiation, pressure induced homogenization, and electrically induced disintegration. Ball milling is the simplest mechanical method for sludge pretreatment [43]. In this method, the larger sized particles are reduced to smaller particles which increases the surface area of the substrates and helps in their easy degradation [44]. The soluble COD also increases due to treatment of sewage sludge by ball milling which in turn increases the biogas production in the subsequent anaerobic digestion step due to the availability of more utilizable organic substrates [45]. However, the requirement of energy to operate ball milling could be detrimental to the cost of biogas production. Furthermore, sometimes the creation of smaller sized particles due to ball milling of sewage sludge can cause rapid formation of acid inside the reactor creating unsuitable environment for methanogens causing reactor failure [46].

The sonication or ultrasonication is another method commonly used in sludge pretreatment for biogas production due to its high effectiveness [47]. The combination of hydro-mechanical shear force created in this method helps in disintegrating the sludge flock structure and the hard cell wall of the microbes present within sludge [48]. The key parameters effecting the ultrasonication performance for sludge pretreatment are the intensity of sonication, temperature, and sludge characteristics. Among these sonication intensity (frequency) and temperature have a positive correlation with their increasing values on the biogas production from the pretreated sludge [46]. Again the requirement of high energy and associated increased cost can render this method unsuitable for large-scale operation which could be the only drawback of this method.

Microwave irradiation is not so common as a method for sludge pretreatment. In this process, sludge is subjected to high intensity microwave radiation which causes sludge flocks and complex biomass structure to disintegrate [49]. The organic matter solubilization also increases in this process which helps in their easy utilization during acidogenesis and following methanogenesis [50]. In addition to the high operating cost of this method for sludge pretreatment, the other disadvantages are related to its non-uniform effect on the biomass during treatment and as it does not degrade biomass directly [51]. In the case of pressure induced homogenization, the sludge is subjected to sudden increased pressure ($<$10000 Psi) which causes severe shear stress on the biomass [52]. This high pressure makes the sludge solubilize and larger size particles within sludge undergo hydrolysis. This helps in better enzymatic reaction on the substrate [53, 54]. However, high energy requirement and related cost are its drawbacks. This method has seldom been used at industrial-scale and despite having promising potential further research is needed for its commercial implementation.

Electrical disintegration involves high-strength electrical pulse mediated sludge treatment during which the larger size flocks break down into smaller particles and also helps in cellular dissolution [55]. This dissolution causes increase in soluble COD and improves the soluble COD to total COD ratio which is an indicator of good biodegradability of the sewage sludge [49]. Although, the biogas production significantly increased due to this pretreatment, the high cost of this process is one of the major hindrances in its commercial application.

3.5.1.2 Chemical Treatment

Chemical pretreatment methods such as alkali treatment, acid treatment, and ozonation are widely used methods for sewage sludge treatment to enhance their biogas production potential. The ease of operation and lower cost compared to the other physical and mechanical methods are the reasons behind their popularity. The process efficiency in terms of increase in biogas production from chemically treated sludge depends on a few factors such as characteristics of the sludge, type of chemical agent used, reaction time, and temperature. The sludge characteristics is one of the most prominent aspect in this regard, as chemical treatment is effective only in the case of high cellulose containing substrates rather than easily biodegradable substrates.

As the name suggests, alkali treatment involves treating the sewage sludge with alkali compounds, namely sodium hydroxide, potassium hydroxide, magnesium hydroxide, and calcium hydroxide. The mechanisms involved in alkali treatment are the cellular lysis and disruption of the extracellular polymeric substances associated with sludge, these cause the organic matter to be solubilized which increases soluble COD of the substrates and helps in their easy utilization. Alkali treatment of sewage sludge can increase the pH of the resulting solution to as high as 10 or 11, but even at such high pH values biogas production could be achieved. However, it is better to reduce the pH to optimum levels for better methanogesis performance which will involve added cost for additional chemical and for a large scale setup may significantly impact the overall process cost. The alkali treatment, if done at a very high concentrations, can leave a significant amount on the substrate which can cause inhibition to biogas production and can potentially destabilize the buffering system inside the bioreactor. The other drawback of alkali treatment is that if the sludge has a considerable amount of lignin then harmful by-products can be produced which can inhibit the methanogens.

Acid based treatment of sewage sludge uses inorganic acids such as hydrochloric acid, sulphuric acid and nitric acid. Acid treatment is suitable for substrates rich in lignin content. Similar to the alkali treatment, acid treatment also solubilizes organic matter and increases soluble COD by breaking down large size molecules, cellular structure, and polymeric substances present in sewage flocs. The pH of the resulting solution dropped to a very low value (2–3), which is not conducive to the methanogens, however methane production had been reported previously in such low pH. In order to bring the pH to a neutral level, it furth requires other chemicals and increases additional costs. Moreover, due to t corrosive nature of the acids, this type pretreatment requires better quality reactor vessels which may incur additional cost.

Ozonation uses strong oxidizing agent ozone to produce hydroxyl radicals which degrade complex substrates to simpler ones which could be easily biodegraded by microbes. The process efficiency depends on the substrate composition and O_3 dosage in addition to experimental conditions such as pH and temperature. Very high effectiveness of ozonation on sludge treatment to produce biogas has been reported in the literature. However, this technology is at its nascent stage particularly for sewage sludge pretreatment, mainly due to its high cost and specialized instrumentation for ozone generation. Another factor is that ozone is not a stable compound which also negatively affects its prospects. Other oxidizing agents such as hydrogen peroxide and peracetic acid can create similar hydroxyl radicals for sewage sludge pretreatment. The process is similar to that described for ozonation where complex organic compounds are made more biodegradable due to reaction with hydroxyl radicals.

3.5.1.3 Thermal Treatment

Thermal pretreatment of sewage sludge involves subjecting the sludge to different temperatures which are mostly higher than ambient temperature (50–200 °C). At elevated temperatures, the cellular structure disintegrates and helps in solubilizing the organic matter present in sewage sludge [56]. In addition, the other recalcitrant compounds undergo hydrolysis which makes them more biodegradable. This way increase in the available organic compounds causes higher biogas yield from sewage sludge. The critical factors influencing the pretreatment process are the temperature and the reaction time used. Temperatures lower than 100 °C are sufficient for municipal sewage sludge, however, higher temperatures than this may be required if more complex and recalcitrant sludge is being processed [57]. In addition, sludge processing using higher temperatures can potentially kill off all the pathogenic microorganisms present in sewage sludge, making the sludge handling, processing and its disposal safe [58].

Apart from elevated temperature, subjecting the sewage sludge to freezing temperature has also yielded higher biogas in the subsequent anaerobic digestion. In this method, the sludge is cooled rapidly to freezing conditions (−20 °C) and later thawed to ambient temperature [59]. This freezing and thawing cycle causes cell lysis and complex organic structures to solubilize. A better solid–liquid separation could also be achieved in this pretreatment method [60]. Although, this process requires external energy input for such a huge change in temperature, natural freezing conditions in colder region could be explored to achieve similar results without any added cost.

3.5.1.4 Biological Treatment

Biological methods are classified under aerobic, anaerobic, and enzymatic methods. These methods are considered to be environmentally sustainable as they do not involve any hazardous/corrosive chemicals or high temperature/pressure processes. However, as the biological methods are slow processes, these types of pretreatment take longer than the other methods mentioned in previous sections. In the case of aerobic treatment, a separate aerobic consortia can be involved for pretreating sewage sludge during which the compounds resistant to anaerobic digestion are degraded to easily biodegradable compounds. This helps the better degradation of the sludge by hydrolysing the complex substrates present in sewage sludge. Sometimes microaerobic or anoxic environments are also created for facultative anaerobic microorganisms. The exposure to an oxygenated environment is occasionally accompanied with elevated temperature or pH adjustment to impart a combined effect on the sludge to yield better results in a shorter reaction time. Thermophilic microbes possess many of the extracellular hydrolytic enzymes which help in the degradation of complex substrates. The aerobic treatment necessitates an additional unit for air supply and in some cases a separate aerobic reactor may also be required which could cause a rise in process cost. Furthermore, it is essential to ensure proper mixing in the aerobic treatment process in order to ensure good utilization of supplied oxygen and appropriate treatment of the sewage sludge.

Anaerobic treatment is aimed at improving biogas production from sewage sludge by properly adjusting the activity of acidogenic and methanogenic microorganisms present in the anaerobic digestion system. This type of pretreatment method involves a two stage process with a significant difference in temperature in the two stages. Commonly the pretreatment stage is the thermophilic one with a short residence time, where degradation of complex substrate takes place. The subsequent methanogenic phase is conducted under moderate temperature (35–45 °C). Proper monitoring and controlling in this type of system is essential to ensure the desired result due to the sensitivity of the microorganisms involved. Hence, skilled personnel with good handling capability is required to run such an operation. In this regard, the additional cost cannot be avoided.

The enzymatic method involves treating the sewage sludge with hydrolytic enzymes to break down the complex compounds present in the sludge. The process can be carried out under different modes: enzymes can be directly added in the anaerobic digestion system or a separate pretreatment step could be designed. If the anaerobic digestion process for biogas production is a two-step one, the addition of enzymes could also be done in the acidogenesis step. Although, enzyme treatment was found to be effective in enhancing biogas production from sewage sludge, many key factors such as enzyme dosage, substrate characteristics, temperature, pH, and reaction time play very important roles in determining the process success. The enzymes are often sensitive and have an optimum condition for their reaction, which should be examined clearly before moving ahead with the pretreatment. Apart from extracted (cell free) enzymes, whole enzyme containing biomass including those of various fungal species have been reported to be effective in pretreating sludge for biogas production.

3.5.2 Effect of Additives

Different additives including micronutrients, trace elements, nanoparticles, and carbonaceous compounds have been examined for enhancing anaerobic digestion of sewage sludge. The micronutrient addition simulates enzyme activity of the microorganism involved in different stages of anaerobic digestion [61]. The different trace metals used in such cases are known to form the core of different enzyme structures and also act as co-factors in many important enzymes [62]. It was also reported that these trace nutrients favour certain types of beneficial microbes and increase their relevance inside the reactor, by which they direct certain specific product formation. The iron, iron nanoparticles or the carbon compounds have additional advantages due to their conducting properties which makes them conducive for interspecies electron transfer. This electron transfer also helps in enhancing the process efficiency and provides a higher yield of methane in a shorter retention time.

The trace elements supplementation has also been found to improve anaerobic digestion performance of sewage sludge. In a study by [63] Fe, Ni, Zn, Cu, Co, and Mo dosage was optimized to enhance methane production for co-digestion of food waste and sewage sludge at various organic loading condition (1.2–18 g L^{-}). The optimum doses for Fe, Ni, Zn, Cu, Co, and Mo were 5000, 200, 320, 250, 3.0, and 2.2 mg L^{-1}, respectively. The authors observed that the addition of trace elements increased biodegradability of the organic compounds and increased acid consumption. The addition of Cu and Co prevented ammonia accumulation inside the bioreactor even at high organic loading conditions which made the biogas production performance more stable.

Jiang et al. [62] studied the effect of the addition of trace elements to enzyme pretreated sewage sludge and found ~ 85% increase in methane production at 1.24% concentration of trace element solution. This trace element solution contained 98.8% iron, 0.6% nickel, 0.2% cobalt, 0.02% tungstate, 0.4% molybdenum, and 0.01% yeast extract. The addition of trace element solution increased the acetogenic and protein degrading microorganisms in addition to hydrogenotrophic methanogenic archaeal population.

Zhang and Wang [64] studied the effect of manganese iron oxide coated biochar on sewage sludge ananerobic digestion. The addition of $MnFe_2O_4$ containing biochar increased the methane production by 56% at its optimum concentration (1.5 g), which was attributed to the enhancement in biodegradability of organic matter including VFA (~35% higher utilization). In addition, due to this compound being a conducting material the interspecies elctron transfer through this was also suggested and was confirmed by the selective enrichment of *Methanosarcina* sp., capable of interspecies electron transfer, in Mn and Fe containing biochar amended samples.

Lizama et al. [65] reported that the addition of zero valent iron nanoparticles enabled enhanced methane production from sewage sludge at 9 mg g^{-1} VSS concentration. The addition of iron nanoparticles increased methane production by more than twice its control experiment and the methane content in the biogas also improved significantly. The mechanism involved in methane production was related to reduction of oxidation-reduction potential due to the addition of iron nanoparticles and subsequent increase in acidogenic degradation of sewage sludge.

Thiruselvi et al. [66] reported iron nanoparticle mediated increase in methane production from sewage sludge. In this work the iron nanoparticles were synthesized using leaf extract of *Vitex negundo*, which when added to the methanogenic reactor at 200 mg L^{-1} concentration produced a very high methane concentration of 0.45 L g^{-1} of VSS. In addition, the COD and solids removal were 86% and 94% which was considerably higher than the values obtained in control experiment without any nanoparticle addition.

Other unconventional materials such as coal derived fly ash was explored for increasing anaerobic digestion of sewage sludge. For example, Montalvo et al. [67] examined the effect of fly ash particle size (0.8–2.36 mm) at a fixed concentration (50 mg L^{-1}) on biogas production, and found that 1.0–1.4 mm size particles have the most significant effect on the methanogenic conversion of sludge. The favourable effect of moderate size particles was attributed to the possible attachment of desirable microbes on the fly ash surface, which was not possible on small particles. The mineral presence in small amounts also had a positive impact as these elements can act as micronutrient and simulate microbial performance. The COD and solids removal efficiencies were also improved significantly with the addition of fly ash. These findings were confirmed by another study where an 18% increase in methane production from winery wastewater was observed by the addition of 25 mg L^{-1} fly ash [68]. However, a high fly ash concentration inhibited the methanogenic activity.

Similar enhancement of anaerobic digestion due to fly ash addition has also been reported in the case of two stage digestion of sewage sludge [69]. The effect of temperature, pH, and reaction time was investigated in this work. The results showed that the hydrolysis was faster at high temperatures (55 and 65 °C), whereas lower temperatures favoured acidogenesis. The addition of ash reduced retention time and simulated sludge hydrolysis. The maximum volatile fatty acid production (10 g L^{-1}) was observed at neutral pH (7.0) and at four days of reaction time.

3.6 Conclusion and Future Research Direction

Anaerobic digestion is a promising bioprocess for treating sewage sludge effectively along with producing value added products such as biogas, volatile fatty acids, and digestate. The important process parameters are solid content, composition of the sewage sludge, pH and temperature of the anaerobic process, the presence of toxic pollutants such as heavy metals and xenobiotics. The bioreactor setup also plays a key role in deciding the treatment efficiency and the biogas yield. The high solid content of sewage sludge is a challenge for its anaerobic digestion. To overcome which various pretreatment technologies are proposed in this study. Moreover, co-digestion strategies where sewage sludge is digested along with organic waste from other industries have shown higher biogas yield and methane content. There have been recent studies on the addition of different materials such as fly ash, biochar, iron nanoparticles, etc. to the anaerobic digestion process to enhance the performance of sewage sludge digestion. There is a great deal of scope in this area of research in exploring novel materials for application in sewage sludge digestion. However, special care should be taken in recovering such materials after the digestion process. This is important not only from the recycling perspective but also for preventing future pollution in the downstream if they escape the digester along with treated effluent. There is also scope for research in the area of bioreactor monitoring and controlling for improving digestion performance. This is essential as the anaerobic digestion process is a sensitive one with possible microbial deactivation at any stage due to changes in environmental parameters (pH, temperature, etc.). Apart from these, there is an opportunity for artificial neural network based modelling work in future for better prediction of reactor performance which should be focused on. Implementation of internet-of-things based data driven technologies in anaerobic digestion could also be taken up as future research objective. This will help this waste management process to transition to the technology age and will be ideally suited for industry 4.0 technologies.

References

1 Sinharoy, A., Pakshirajan, K., and Lens, P.N.L. (2020a). Biological sulfate reduction using gaseous substrates to treat acid mine drainage. *Current Pollution Reports* 6: 328–344.

2 Mahesh, R., Panda, S.K., Das, M. et al. (2021). Advances in biotechnological tools for bioremediation of wastewater using bacterial–algal symbiotic system. In: *Wastewater Treatment: Cutting-Edge Molecular Tools, Techniques and Applied Aspects* (ed. M.P. Shah, A. Sarkar, S. Mandal), 385–411. Netherlands: Elsevier. https//doi.org/10.1016/B978-0-12-821881-5.00019-2.

3 Sinharoy, A., Kumar, M., and Pakshirajan, K. (2020b). An overview of bioreactor configurations and operational strategies for dark fermentative biohydrogen production. *Bioreactors* 249-288: doi: 10.1016/B978-0-12-821264-6.00014-0.

4 Fytili, D. and Zabaniotou, A. (2008). Utilization of sewage sludge in EU application of old and new methods—a review. *Renewable and Sustainable Energy Reviews* 12 (1): 116–140.

5 Kirchmann, H., Börjesson, G., Kätterer, T., and Cohen, Y. (2017). From agricultural use of sewage sludge to nutrient extraction: a soil science outlook. *Ambio* 46 (2): 143–154.

6 Sinharoy, A., Baskaran, D., and Pakshirajan, K. (2019a). Sustainable biohydrogen production by dark fermentation using carbon monoxide as the sole carbon and energy source. *International Journal of Hydrogen Energy* 44 (26): 13114–13125.

7 Sinharoy, A., Baskaran, D., and Pakshirajan, K. (2019b). A novel carbon monoxide fed moving bed biofilm reactor for sulfate rich wastewater treatment. *Journal of Environmental Management* 249: 109402.

8 Beniche, I., Hungría, J., El Bari, H. et al. (2021). Effects of C/N ratio on anaerobic co-digestion of cabbage, cauliflower, and restaurant food waste. *Biomass Conversion and Biorefinery* 11 (5): 2133–2145.

9 Varjani, S.J., Gnansounou, E., Baskar, G. Pant, D., and Zakaria, Z.A. (2018). Introduction to waste bioremediation. In: *Waste Bioremediation, Energy, Environment, and Sustainability* (ed. S. Varjani, E. Gnansounou, B. Gurunathan et al.), 1–5. Singapore: Springer. doi: 10.1007/978-981-10-7413-4_1.

10 Wang, X., Lu, X., Li, F., and Yang, G. (2014). Effects of temperature and carbon-nitrogen (C/N) ratio on the performance of anaerobic co-digestion of dairy manure, chicken manure and rice straw: focusing on ammonia inhibition. *Plos one* 9 (5): 1–7.

11 Tomei, M.C., Braguglia, C.M., Cento, G., and Mininni, G. (2009). Modeling of anaerobic digestion of sludge. *Critical Reviews in Environmental Science and Technology* 39 (12): 1003–1051.

12 Ho, D., Jensen, P., and Batstone, D. (2014). Effects of temperature and hydraulic retention time on acetotrophic pathways and performance in high-rate sludge digestion. *Environmental Science and Technology* 48 (11): 6468–6476.

13 Rivière, D., Desvignes, V., Pelletier, E. et al. (2009). Towards the definition of a core of microorganisms involved in anaerobic digestion of sludge. *ISME Journal* 3 (6): 700–714.

14 Kumar, M., Dutta, S., You, S. et al. (2021). A critical review on biochar for enhancing biogas production from anaerobic digestion of food waste and sludge. *Journal of Cleaner Production* 305: 127143.

15 Mata-Alvarez, J., Macé, S., and Llabres, P. (2000). Anaerobic digestion of organic solid wastes. *An Overview of Research Achievements and Perspectives. Bioresource Technology* 74 (1): 3–16.

16 Adekunle, K.F. and Okolie, J.A. (2015). A review of biochemical process of anaerobic digestion. *Advances in Bioscience and Biotechnology* 06 (03): 205–212.

17 Sinharoy, A. and Pakshirajan, K. (2021). Methane free biohydrogen production from carbon monoxide using a continuously operated moving bed biofilm reactor. *International Journal of Hydrogen Energy* 46 (1): 306–313.

18 Paul, T., Sinharoy, A., Baskaran, D. et al. (2020b). Bio-oil production from oleaginous microorganisms using hydrothermal liquefaction: a biorefinery approach. *Critical Reviews in Environmental Science and Technology* doi: 10.1080/10643389.2020.1820803.

19 Paul, T., Sinharoy, A., Pakshirajan, K., and Pugazhenthi, G. (2020a). Lipid-rich bacterial biomass production using refinery wastewater in a bubble column bioreactor for bio-oil conversion by hydrothermal liquefaction. *Journal of Water Process Engineering* 37: 101462.

20 Schnell, M., Horst, T., and Quicker, P. (2020). Thermal treatment of sewage sludge in Germany: a review. *Journal of Environmental Management* 263: 110367.

21 Chen, Y., Cheng, J.J., and Creamer, K.S. (2008). Inhibition of anaerobic digestion process: a review. *Bioresource Technology* 99 (10): 4044–4064.

22 Bowen, E.J., Dolfing, J., Davenport, R.J., Read, F.L., and Curtis, T.P. (2014). Low-temperature limitation of bioreactor sludge in anaerobic treatment of domestic wastewater. *Water Science and Technology* 69 (5): 1004–1013.

23 Mao, C., Feng, Y., Wang, X., and Ren, G. (2015). Review on research achievements of biogas from anaerobic digestion. *Renewable and Sustainable Energy Reviews* 45: 540–555.

24 Qin, S., Wainaina, S., Liu, H. et al. (2021). Microbial dynamics during anaerobic digestion of sewage sludge combined with food waste at high organic loading rates in immersed membrane bioreactors. *Fuel* 303: 121276.

25 Zhao, J., Li, Y., Pan, S., Tu, Q., and Zhu, H. (2019). Performance of a forward osmotic membrane bioreactor for anaerobic digestion of waste sludge with increasing solid concentration. *Journal of Environmental Management* 246: 239–246.

26 Li, Y., Song, Z., Yuan, Y., Zhang, Q., and Zhu, H. (2020). Rheology improvement in an osmotic membrane bioreactor for waste sludge anaerobic digestion and the implication on agitation energy consumption. *Bioresource Technology* 295: 122313.

27 Wandera, S.M., Qiao, W., Jiang, M. et al. (2019). Enhanced methanization of sewage sludge using an anaerobic membrane bioreactor integrated with hyperthermophilic biological hydrolysis. *Energy Conversion and Management* 196: 846–855.

28 Niu, C., Pan, Y., Lu, X. et al. (2020). Mesophilic anaerobic digestion of thermally hydrolyzed sludge in anaerobic membrane bioreactor: long-term performance, microbial community dynamics and membrane fouling mitigation. *Journal of Membrane Science* 612: 118264.

29 Lopez, A., Rodriguez-Chueca, J., Mosteo, R., Gomez, J., and Ormad, M.P. (2020). Microbiological quality of sewage sludge after digestion treatment: a pilot scale case of study. *Journal of Cleaner Production* 254: 120101.

30 Zupančič, G.D., Uranjek-Ževart, N., and Roš, M. (2008). Full-scale anaerobic co-digestion of organic waste and municipal sludge. *Biomass and Bioenergy* 32 (2): 162–167.

31 Maragkaki, A.E., Fountoulakis, M., Gypakis, A. et al. (2017). Pilot-scale anaerobic co-digestion of sewage sludge with agro-industrial by-products for increased biogas production of existing digesters at wastewater treatment plants. *Waste Management* 59: 362–370.

32 Liu, X., Gao, X., Wang, W. et al. (2012). Pilot-scale anaerobic co-digestion of municipal biomass waste: focusing on biogas production and GHG reduction. *Renewable Energy* 44: 463–468.

33 Li, C., Wang, X., Zhang, G. et al. (2018). A process combining hydrothermal pretreatment, anaerobic digestion and pyrolysis for sewage sludge dewatering and co-production of biogas and biochar: pilot-scale verification. *Bioresource Technology* 254: 187–193.

34 Li, C., Wang, X., Zhang, G. et al. (2017). Hydrothermal and alkaline hydrothermal pretreatments plus anaerobic digestion of sewage sludge for dewatering and biogas production: bench-scale research and pilot-scale verification. *Water Research* 117: 49–57.

35 Lloret, E., Pastor, L., Pradas, P., and Pascual, J.A. (2013). Semi full-scale thermophilic anaerobic digestion (TAnD) for advanced treatment of sewage sludge: stabilization process and pathogen reduction. *Chemical Engineering Journal* 232: 42–50.

36 Rintala, J.A. and Järvinen, K.T. (1996). Full-scale mesophilic anaerobic co-digestion of municipal solid waste and sewage sludge: methane production characteristics. *Waste Management & Research* 14 (2): 163–170.

37 Liu, J. and Smith, S.R. (2020). A multi-level biogas model to optimise the energy balance of full-scale sewage sludge conventional and THP anaerobic digestion. *Renewable Energy* 159: 756–766.

38 Lee, J., Kim, E., Han, G. et al. (2018). Microbial communities underpinning mesophilic anaerobic digesters treating food wastewater or sewage sludge: a full-scale study. *Bioresource Technology* 259: 388–397.

39 Shin, S.G., Koo, T., Lee, J. et al. (2016). Correlations between bacterial populations and process parameters in four full-scale anaerobic digesters treating sewage sludge. *Bioresource Technology* 214: 711–721.

40 Wang, P., Yu, Z., Zhao, J., and Zhang, H. (2018). Do microbial communities in an anaerobic bioreactor change with continuous feeding sludge into a full-scale anaerobic digestion system? *Bioresource Technology* 249: 89–98.

41 Kim, J., Kim, W., and Lee, C. (2013). Absolute dominance of hydrogenotrophic methanogens in full-scale anaerobic sewage sludge digesters. *Journal of Environmental Sciences* 25 (11): 2272–2280.

42 Jenicek, P., Bartacek, J., Kutil, J., Zabranska, J., and Dohanyos, M. (2012). Potentials and limits of anaerobic digestion of sewage sludge: energy self-sufficient municipal wastewater treatment plant? *Water Science & Technology* 66 (6): 1277–1281.

43 Zhu, J.Y., Wang, G.S., Pan, X.J., and Gleisner, R. (2009). Specific surface to evaluate the efficiencies of milling and pretreatment of wood for enzymatic saccharification. *Chemical Engineering Science* 64 (3): 474–485.

44 Kopp, J., Miiller, J., Dichtl, N., and Schwedes, J. (1997). Anaerobic digestion and dewatering characteristics of mechanically disintegrated excess sludge. *Water Science and Technology* 36 (11): 129–136.

45 Baier, U. and Schmidheiny, P. (1997). Enhanced anaerobic degradation of mechanically disintegrated sludge. *Water Science and Technology* 36 (11): 137–143.

46 Nguyen, V.K., Chaudhary, D.K., Dahal, R.H. et al. (2021). Review on pretreatment techniques to improve anaerobic digestion of sewage sludge. *Fuel* 285: 119105.

47 Kavitha, S., Kannah, R.Y., Gunasekaran, M. et al. (2019). Effect of low intensity sonic mediated fragmentation of anaerobic granules on biosurfactant secreting bacterial pretreatment: energy and mass balance analysis. *Bioresource Technology* 279: 156–165.

48 Atelge, M.R., Atabani, A.E., Banu, J.R. et al. (2020). A critical review of pretreatment technologies to enhance anaerobic digestion and energy recovery. *Fuel* 270: 117494.

49 Lee, I.S. and Rittmann, B.E. (2011). Effect of low solids retention time and focused pulsed pre-treatment on anaerobic digestion of waste activated sludge. *Bioresource Technology* 102 (3): 2542–2548.

50 Choi, H., Jeong, S.W., and Chung, Y.J. (2006). Enhanced anaerobic gas production of waste activated sludge pretreated by pulse power technique. *Bioresource Technology* 97 (2): 198–203.

51 Beszédes, S., László, Z., Szabó, G., and Hodúr, C. (2011). Effects of microwave pretreatments on the anaerobic digestion of food industrial sewage sludge. *Environmental Progress & Sustainable Energy* 30 (3): 486–492.

52 Nabi, M., Zhang, G., Zhang, P. et al. (2019). Contribution of solid and liquid fractions of sewage sludge pretreated by high pressure homogenization to biogas production. *Bioresource Technology* 286: 121378.

53 Salihu, A. and Alam, M.Z. (2016). Pretreatment methods of organic wastes for biogas production. *Journal of Applied Sciences* 16 (3): 124–137.

54 Zhang, Y., Zhang, P., Guo, J. et al. (2013). Sewage sludge solubilization by high-pressure homogenization. *Water Science and Technology* 67 (11): 2399–2405.

55 Neumann, P., Pesante, S., Venegas, M., and Vidal, G. (2016). Developments in pre-treatment methods to improve anaerobic digestion of sewage sludge. *Reviews in Environmental Science and Bio/ Technology* 15 (2): 173–211.
56 Jo, H.B., Parker, W., and Kianmehr, P. (2018). Comparison of the impacts of thermal pretreatment on waste activated sludge using aerobic and anaerobic digestion. *Water Science and Technology* 78 (8): 1772–1781.
57 Nazari, L., Yuan, Z., Santoro, D. et al. (2017). Low-temperature thermal pre-treatment of municipal wastewater sludge: process optimization and effects on solubilization and anaerobic degradation. *Water Research* 113: 111–123.
58 De Los Cobos-vasconcelos, D., Villalba-Pastrana, M.E., and Noyola, A. (2015). Effective pathogen removal by low temperature thermal pre-treatment and anaerobic digestion for class A biosolids production from sewage sludge. *Journal of WaterSanitation and Hygiene for Development* 5 (1): 56–63.
59 Montusiewicz, A., Lebiocka, M., Rożej, A., Zacharska, E., and Pawłowski, L. (2010). Freezing/thawing effects on anaerobic digestion of mixed sewage sludge. *Bioresource Technology* 101 (10): 3466–3473.
60 Hu, K., Jiang, J.Q., Zhao, Q.L. et al. (2011). Conditioning of wastewater sludge using freezing and thawing: role of curing. *Water Research* 45 (18): 5969–5976.
61 Hijazi, O., Abdelsalam, E., Samer, M. et al. (2020). Environmental impacts concerning the addition of trace metals in the process of biogas production from anaerobic digestion of slurry. *Journal of Cleaner Production* 243: 118593.
62 Jiang, X., Lyu, Q., Bi, L. et al. (2022). Improvement of sewage sludge anaerobic digestion through synergistic effect combined trace elements enhancer with enzyme pretreatment and microbial community response. *Chemosphere* 286: 131356.
63 Bardi, M.J. and Aminirad, H. (2020). Synergistic effects of co-trace elements on anaerobic co-digestion of food waste and sewage sludge at high organic load. *Environmental Science & Pollution Research* 27 (15).
64 Zhang, M. and Wang, Y. (2020). Effects of Fe-Mn-modified biochar addition on anaerobic digestion of sewage sludge: biomethane production, heavy metal speciation and performance stability. *Bioresource Technology* 313: 123695.
65 Lizama, A.C., Figueiras, C.C., Pedreguera, A.Z., and Espinoza, J.E.R. (2019). Enhancing the performance and stability of the anaerobic digestion of sewage sludge by zero valent iron nanoparticles dosage. *Bioresource Technology* 275: 352–359.
66 Thiruselvi, D., Yuvarani, M., Salma, A. et al. (2021). Enhanced biogas from sewage sludge digestion using iron nanocatalyst from vitex negundo leaf extract: response surface modeling. *International Journal of Environmental Science and Technology* 18 (8): 2161–2172.

67 Montalvo, S., Cahn, I., Borja, R., Huiliñir, C., and Guerrero, L. (2017). Use of solid residue from thermal power plant (fly ash) for enhancing sewage sludge anaerobic digestion: influence of fly ash particle size. *Bioresource Technology* 244: 416–422.

68 Lauzurique, Y., Fermoso, F.G., Sánchez, N. et al. (2021). Effect of the addition of fly ash on the specific methane production and microbial communities in the anaerobic digestion of real winery wastewater. *Journal of Chemical Technology & Biotechnology* 96 (10): 2882–2890.

69 Niero, L., Morgan-Sagastume, F., and Lagerkvist, A. (2021). Accelerating acidogenic fermentation of sewage sludge with ash addition. *Journal of Environmental Chemical Engineering* 9 (6): 106564.

4

Potential of Treated Industrial Effluent – Challenges and Opportunities

Vaishnavi Sharma[1] *and Satarupa Dey*[2,*]

[1] *National Institute of Technology Warangal, Telangana, India*
[2] *Department of Botany, Shyampur Siddheswari Mahavidyalaya, Howrah. West Bengal, India*
* *Corresponding author*

4.1 Introduction

The circular economy concept has been flourishing in the last few decades. Its application in various activities has boomed. The application of this concept in the industrial sector will involve noteworthy collaboration from the industries of used water as a preferable water source. The hurdles in the application of this concept to the industrial sector are many, e.g., people's viewpoint on it, economic considerations, lack of governmental control and policies etc. [1]. Considering the waste management hierarchy, recycling and reuse of industrial wastewater could be a pioneering practice to bring the resource back into the loop and hence close it in a sustainable way. The alliance of the sustainability concept and the circular economy concept to deal with wastewater coming from industry will have various advantages. Once the treated water reaches the secure standards for a particular use, it can be widely used; this will ensure less use of virgin water and will also positively impact the environment [1]. Water being a scarce resource needs to be used more judicially. Increasing urbanization has led to the development of better living standards which now poses demands for better municipal supply systems and sanitation which are also contributing factors in escalated water usage. Furthermore, the changing climatic conditions has generated a gap between supply and demand. This finally produced a need for better approaches to deal with the water crisis worldwide [1]. The World Economic Forum Global Risk Report stated a 40% divergence in the supply of water and demand by the year 2030 if the current practices are continued [1]. The industrial revolution involved the increased use of the gifts of nature [2]. Altogether, industry uses about 19% of the world's water. Asia uses about 10% and Europe about 57% of the entire water use

Application of Sewage Sludge in Industrial Wastewater Treatment, First Edition. Edited by Maulin P. Shah.

[1]. And the increased pace of the growth will definitely come with more water usage by industry[2]. The agricultural sector is found to be using the greatest percentage of the water resources [3]. Although the next in line are industries which are the major sources of pollution. Some of the sources are: (a) pulp and paper industries, coal utilizing industries use very high volumes of water, (b) leather industries are often found to be using toxic chemicals which are discharged into streams after treatment, (c) electroplating industries require extremely good quality water for operating purposes but mostly render different pollutants in the final substrate. Different industries use different quality and quantities of water for different operations. All industries also have varied capabilities of reusing such treated wastewater. A few of the industries are mentioned in the Table 4.1 [2]:

Table 4.1 Approximate capacity of industries to reuse treated industrial wastewater [2].

High Capacity	Medium Capacity	Low Capacity
• Pulp and Paper • Cotton Textile • Glass and Steel	• Slaughterhouse • Dairy • Canning and Food Processing • Distillery • Wool Textile • Photographic Processing • Chemical • Fertilizer • Oil Refining • Petroleum • Meat Processing • Electroplating	• Tanneries and Leather finishing • Pesticide • Rubber • Aluminium • Explosives Manufacturing • Paint Manufacturing

The main task is the identification of industries with maximum potential and maximum water usage to decide the requirement of reuse of treated industrial water [2].

Economic expansion is often accompanied by extra water usage. Recycling and re-using the same water from industries is a global aspiration. The simplest approach in this regard can be the separation of liquid and solid components in the wastewater. For this, various physical, chemical and biological methods are brought into use as explained in Fig 4.1. The easy physical methods used for this can be plain sedimentation and other methods [2]. There are many technologies involved in reaching the goal of closing the water loop, such as, desalination, micro-filtration followed by reverse osmosis [1], micro-filtration, ultra-filtration, activated carbon, ion-exchange methods, chemical coagulation and precipitation methods [2], phytoremediation [4] etc. Desalination, however, has disadvantages on the economic and environmental scale, as the brine released from the industry requires a series of treatments before disposal or reaching the required re-use standards. Pathogens and remaining organic nutrients often find their way out of reverse osmosis and filtration procedures. So these advanced treatments fail to treat the water to secure its specific use [1]. Conventional pre-treatment and secondary treatment processes cannot produce treated water which can be conveniently reused again. Capable conventional treatments to ensure water reuse can be shown in the following figure.

Primary treatments have following features [5]:

- Energy is needed for mixing and/or oxygen supply.
- Important civil works are involved (concrete, aeration and stirring systems, electric power wiring, pipelines, internal roads, etc.).
- Less surface occupied by the units.
- The processes can be quickly influenced using extra energy.
- Economy: high costs of energy and by-product management, comparatively reduced cost of land.
- Operation and maintenance by highly skilled (trained staff). High amounts of by-product, economy of scale exists.

Secondary treatments have the following features in comparison to the primary ones [5]:

- Soft energy is needed (sun, wind, etc.). External energy used only for pumping water or illumination of the facility. Minimal use of energy from the network.
- Limited civil works are involved: earth moving, water pipelines, access and internal roads.
- Greater space occupied by the units.
- Economy: nearly no cost of energy, comparatively high cost of land.
- Reduced operation and maintenance: the staff must know the characteristics of the processes used.

All these technologies are orthodox and can no longer be applied to reach the required standards. So, mixed treatment technologies involving various physical, chemical and biological processes must be adopted for treatment of industrial wastewater. To ensure the re-use of treated industrial wastewater, tertiary treatment is compulsorily provided to the substrate. Based on the need of a particular quality of treated water, processes are adopted for the effective removal of contaminants like microbes, pathogens, bacteria, heavy metals, nutrients, etc. [1]. Although adoption of expensive treatment technologies without sufficient knowledge of the purpose of the treated wastewater can lead to unnecessary waste of money. Hence, there has to a thorough risk assessment of any scheme which will

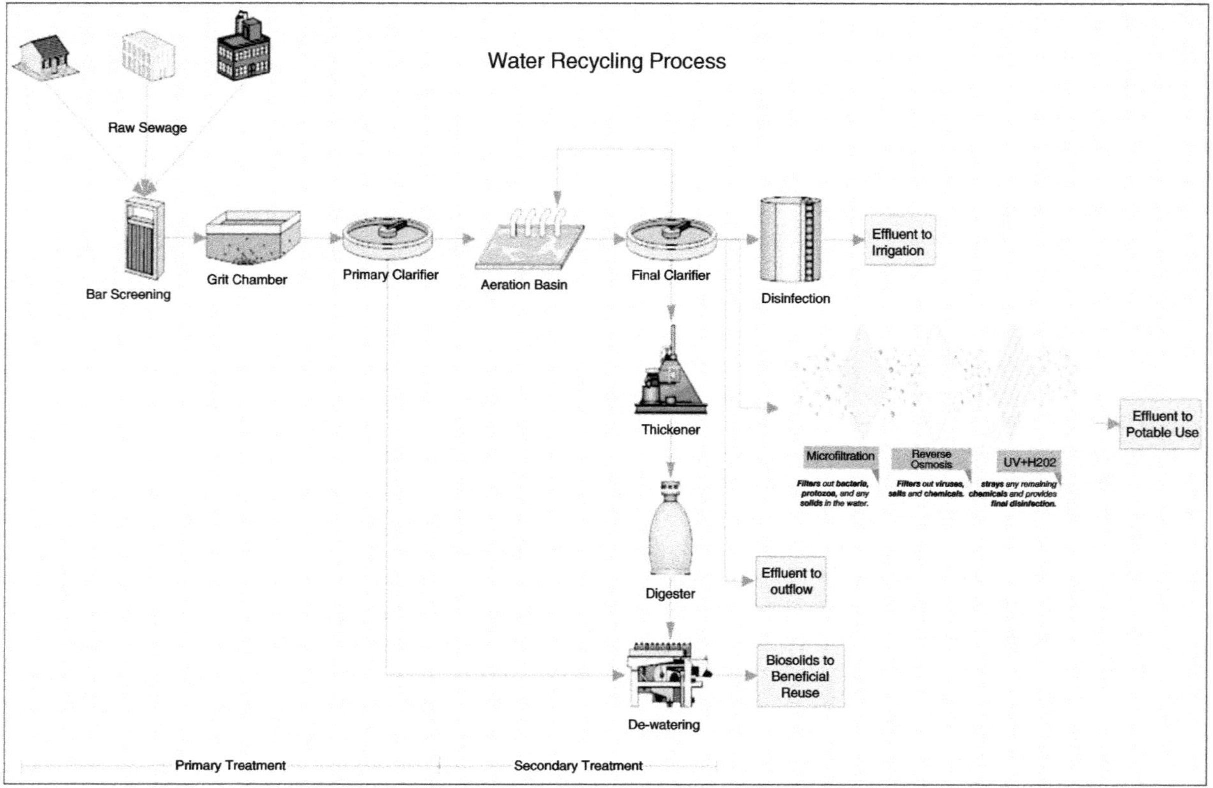

Figure 4.1 Conventional biological wastewater treatment and possible reuse of wastewater [1].

present a comparative study of the benefits or drawbacks involved in adopting any set of operations [1]. Hence the water reuse potential of any industry can be scaled on the basis of the volume of waste created, concentration and composition of different pollutants in it, treatment technologies being used presently, the operational cost of all the units, maintenance cost of the plant, availability of an alternative source of water, availability of buyers of the treated wastewater and lastly the disposal standards specified by the authorities [2].

4.2 Physio-chemical Characteristics of Industrial Water

4.2.1 Characteristics of Industrial Influent and Effluent

It is important for industries to incorporate better technologies to reuse the wastewater they produce. The process of discharging the water after use is more expensive for the industries as the discharge standards are strict and hard to meet for most industries [2]. The discharge quality standard of treated industrial waste water and agricultural reuse water is represented in Table 4.2, 4.3 and 4.4. Different industries can opt for different process chains in order to effectively use the resources again. A beneficial way of treating these waters is to segregate the components from different units mixed in the water from a specific process. Treat this water up to a certain level and incorporate its use in another process. This way the discharge standards of the effluents can be easily achieved [2]. The water reused from the industries find their place in four sectors mostly [6]; (a) agricultural irrigation purposes, (b) groundwater restoration and recharge of aquifers, (c) water required for industries for different processes, and (d) combined usage.

Table 4.2 Discharge quality standards of treated industrial wastewater [3].

Constituent	Standard (mg L^{-1})	Constituent	Standard (mg L^{-1})	Constituent	Standard (mg L^{-1})
Total solid	<500	Iron	<0.5	Molybdenum	<0.07
Dissolved oxygen	>6	Manganese	<0.2	Nickel	<0.02
pH	6.5–8.5	Copper	<0.01	Aldrin and dieldrin	<0.00003
Biological oxygen demand (BOD)	<6	Zinc	<0.01	Alachlor	<0.02
Chemical oxygen demand (COD)	<10	Fluorides	<0.5	Aldicarb	<0.01

(Continued)

Table 4.2 (Continued)

Constituent	Standard (mg L^{-1})	Constituent	Standard (mg L^{-1})	Constituent	Standard (mg L^{-1})
Organic-N	<1	Phenol	<0.002	Atrazine	<0.002
Total-N	<3.5	Arsenic	<0.01	Bentazone	<0.03
Total-P	<2	Cadmium	<0.001	Carbofuran	<0.007
NH_3	<0.5	Chromium	<0.05	Chlordane	<0.0002
NO_3	<2	Cyanide	<0.005	2,4 Dichloroprop	<0.03
Fat/oil	<0.1	Lead	<0.01	Fenoprop	<0.009
SO_4	<200	Selenium	<0.01	Mecoprop	<0.01
Mercury	<0.001	Boron	<0.5	2,4,5-T	<0.009

Table 4.3 Discharge standards of treated wastewater for agricultural reuse [3].

Constituent	Standard (mg L^{-1})	Constituent	Standard (mg L^{-1})	Constituent	Standard (mg L^{-1})
Total solid	<1,000	Copper	<1	Aldrin and dieldrin	<0.003
Dissolved oxygen	>5	Zinc	<2	Alachlor	<0.02
pH	6.5–8.5	Phenol	<0.05	Aldicarb	<0.1
Biological oxygen demand (BOD)	<30	Arsenic	<0.01	Atrazine	<0.02
Chemical oxygen demand (COD)	<50	Cadmium	<0.03	Bentazone	<0.3
Total-N	15	Chromium	<0.05	Carbofuran	<0.07
Total-P	3	Cyanide	<0.01	Chlordane	<0.002
Fat/oil	<3	Lead	<0.1	2,4- Dichloroprop	<0.3
Mercury	<0.001	Nickel	0.1	Fenoprop	<0.09
Iron	<3	Selenium	0.01	Mecoprop	<0.1
Manganese	<2	Total coli group in 100 cm^3	5,000	2,4,5-T	<0.09
Temperature	<3 °C Over	Irrigation	Water		

Table 4.4 Discharge standards for mixing of treated industrial wastewater to non-fresh surface water sources [3].

Constituent	Maximum Limit	Constituent	Maximum Limit	Constituent	Maximum Limit
Temperature	<3 °C Over the recipient waterway	NH_3	–	Total coli group in 100 cm^3	1,000
pH	6–9	Phenol	0.05 mg L^{-1}	Alachlor	<0.1 mg L^{-1}
BOD	60 mg L^{-1}	Mercury	0.01 mg L^{-1}	Aldicarb	<0.5 mg L^{-1}
COD	80 mg L^{-1}	Lead	0.1 mg L^{-1}	Atrazine	<0.1 mg L^{-1}
DO	>4 mg L^{-1}	Cadmium	0.003 mg L^{-1}	Bentazone	<0.15 mg L^{-1}
Fat/oil	10 mg L^{-1}	Arsenic	0.05 mg L^{-1}	Carbofuran	<0.35 mg L^{-1}
Total dissolved solids (mg L^{-1})	<2,000	Selenium	0.1 mg L^{-1}	Aldrin and dieldrin	<0.015 mg L^{-1}
Suspended solids	50 mg L^{-1}	Chromium	0.1 mg L^{-1}	Chlordane	<0.01 mg L^{-1}
H_2S	1 mg L^{-1}	Copper	0.5 mg L^{-1}	2,4-Dichloroprop	<0.5 mg L^{-1}
Cyanide	0.1 mg L^{-1}	Nickel	0.5 mg L^{-1}	Fenoprop	<0.45 mg L^{-1}
Total-P	–	Zinc	2 mg L^{-1}	Mecoprop	<0.5 mg L^{-1}
Total-N	–	Iron	1 mg L^{-1}	2,4,5-T	<0.45 mg L^{-1}

4.2.2 Quality of Treated Water

Global water usage has increased which is mostly due to manufacturing processes from various industries [1]. The quality of influent water is a very important aspect to decide on which treatment processes should be carried out according to the demand for particular re-use. Taking the case of Egypt, prolonged application of treated wastewater on its sandy soil, rendered the soil uncountable benefits. It not only changed its texture but also increased the soil's salinity and solidity, water holding capacity, nutrient availability, aggregate stability and infiltration capacity, etc. [3]. Although it is necessary for users to take proper care while dealing with treated wastewater, because it can have components harmful to human health. Even low quality treated water should not be brought into use because it can have negative impact on the users and the environment. Hence, there is a need for proper monitoring systems, to maintain checks on quality standards of water being brought in to use again [3]. Treated effluents can usually be used in the manner discussed below [7]:

a) Evaporation is carried out through artificial impoundments when material recovery is required.
b) Ground water replenishment is carried out through infiltration techniques but there is a risk of pollution of preserved water due to external components.
c) Irrigation of particular crops can be done with treated water. The water can also be used to maintain the greenery of golf courses, public parks and can also be applied to highway surfaces.
d) Artificial lakes can be filled with such water to promote leisure activities like boating, fishing, swimming, water sports, etc.
e) Aquaculture can be carried out in a controlled and monitored manner for commercial purposes.
f) After being treated to specified levels, treated wastewater can find use even in municipal supplies.
g) Circulation of the same water in different industries can also be a good way of reutilizing such treated water.
h) Easy disposal of such treated water into natural sources can also be done if no utilization is done after the treatment.

4.3 Potential Reuse of Treated Effluent

4.3.1 End Use of Wastewater

Joint efforts are required to achieve the 2030 Agenda for Sustainable Development. Global water re-use potential was evaluated to be about 33.7 GL per day to 54.5 GL per day in 2015; the maximum contribution here was by China, the US, North Africa, the Middle East, South Asian countries and Western Europe (Figure 4.2, 4.3 and 4.4) [1]. A third industrial revolution is just around the corner which is well equipped with the circular economy and sustainability concepts. This transition will not only improve the quality of the environment but will also support nations financially [1]. There are many authorities worldwide which have different guidelines to ensure good end use of treated wastewater. WHO have suggested Stockholm Framework for reuse of treated waste in aquaculture which presents proper health risk assessments [3]. Many studies are found to be using the words: "reuse" for using the same water from the operation units itself, "recycling" for reusing the same water after minimum treatment in the same unit and "reconditioning" for treating the substrate to specified standards for particular use [8]. The sustainability of reusing a particular treated industrial wastewater for a particular application depends on many factors. Industries are found to be using less water for operation and maximum water is created as wastewater. So, once reclaimed, water can be brought into use for various purposes [2]. The most common use of treated wastewater is found in cooling feeders and boilers of different units. The problems associated with this practice are scaling, corrosion and bacterial foaming of the units internally. To tackle these issues different treatments like thermal treatment, ozonation, etc. are carried out [2]. Hence, proper economic evaluation needs to be done before the any operation sets are chosen.

4.3.2 Standards

There has been a great deal of variance in the adoption of water reuse guidelines by different countries around the world (Tables 4.5 and 4.6). The rules written and adopted were created decades back and need immediate upgrading. With the new technologies and increasing environmental concerns it is a critical need to upgrade to better regulations worldwide [1].

4.3.3 Benefits Involved

The reuse of wastewater will offer a shift in the linear human cycle to the circular one which will secure sustainability of water resources [1].

Re-use of industrial wastewater has various advantages. The water once treated can be found to be rich in varied nutrients which can be applied for specific purposes. Industrial wastewater once treated to a certain level can be utilized to recharge the water supplies and can also maintain a balance between demand and supply of water. Other merits can be better agricultural products, less energy utilization involved in the production practices, a positive environmental effect, lower nutrient loads due to the use of treated water, etc. [1]. In the search for cheaper alternatives and to fight water scarcity issues, many cities in the world have come up with their own plans to implement a circular economy concept to reuse industrial wastewater. Some examples are given in the figure [1].

In Europe, the Urban Wastewater Treatment Directives (91–276-EEC) has brought a revolution in wastewater reuse trends. The treated water is used for various purposes and is also given advanced treatments for higher quality products [1]. In Egypt, the texture of the sandy soil effectively changed due to the continuous application of treated industrial wastewater. The soil became rich in organic nutrients, its cation exchange capacity improved and hence was able to produce better crop yields for years. This practice not only reduced the gap between the demand and supply of potable water in the region but also improved the quality of land for a long time [3]. Aquatic life is also positively affected with such practices. Fishery industries can have better business with the incorporation of sustainable practices of using treated wastewater. The increasing populations in coastal regions have increased food demand and ensuring safe food is a necessity. The main factors to be kept in check to ensure better fish production are temperature, pH, DO, dissolved gases concentration, levels of ammonia, nitrates and nitrites concentration in water [3].

Table 4.5 Treatment technologies for reuse of water given by different guideline providers around the world [1].

Water Use		Different	Guidelines	Providers			
	USA (EPA)	**Australian guidelines for water recycling**	**EU Guidelines**	**Mediterranean (UNEP)**	**California**	**Washingt-on**	**WHO Guidelines**
Head-on contact with food or food related activities involving public	For Unprocessed Food Crops; Secondary Treatment (ST): Filtration, Disinfection, Ozonation For Processed Food Crops; ST: Secondary Filtration and Disinfection	Household Use: UV-light, MF, Coagulation Municipal Use: ST: Coagulation Filtration Disinfection MF UV-light	Class A: ST (advanced treatments) Class B: ST and Disinfection Class C: ST and Disinfection	Category I: ST+ Filtration+ Disinfection Category II: ST+ Filtration or ST+ Maturation Ponds for infiltration and percolation	Non restricted recreational impoundments: ST, Coagulation, Clarification, Filtration, Disinfection Food crops and urban use: ST, Coagulation, Filtration, Disinfection	Restricted urban reuse: Oxidation, Disinfection Agricultural reuse: Oxidation, Coagulation Filtration, Disinfection	Agricultural reuse: Treatment Processes Commercial reuse: Production Processes Local reuse:
No straight contact with food or food related activities involving public	Non-food produces: Secondary Disinfection	Municipal Use: Secondary treatment and Disinfection	Class D: ST and Storage, Stabilization Ponds or constructed Wetlands	Category III: ST+ Storage or Oxidation ponds Category IV: Pre-treatment given which should not be less than primary sedimentation	Food Crops: ST+ Disinfection Pasture, Coliforms: ST Irrigation of fodder: ST	Agricultural and Industrial reuse: Oxidation and Disinfection	Correct labelling like grey, waste, reclaimed, green or drinking water

Table 4.6 Standard limits of contaminants specified by different international authorities [1].

Water Use	Different	Guideline	Providers				
Head-on contact with food or food related activities involving public	**USA (EPA)** pH – 6–9 test weekly BOD – 10 mg L^{-1} test weekly Turbidity – 2 NTU test continuous Coliform – No detectable test daily CL_2 residual – 1 mg L^{-1} residual minimum test continuous	**Australia** E. coli Not detected in 100 Ml Monitored weekly Turbidity <1 NTU (95%) <5 NTU (max) Monitored continuously online pH 6.5–8.5 Monitored continuously online Disinfection Chlorine: 0.2–2.0 mg L^{-1} residual	**EU Guidelines** Class A: food consumed raw E. coli – 10 colony forming units per 100 ml Test once a week BOD – 10 mg L^{-1} Test once a week TSS – 10 mg L^{-1} Test once a week Turbidity – 5NTU Test daily	**Mediterranean (UNEP)** Category I: Intestinal nematode 0.1 eggs per litre – fortnightly Faecal Coliforms or Escherichia coli 200 colony forming units per 100 ml – twice weekly Physical/Chemical suspended solids 10 mg L^{-1} – weekly	**California** Non restricted Recreational impoundments and food crops and urban use: coliform limits: 2.2/100 mL 23/100 mL in more than one sample in any 30-day period 240 100 mL^{-1}	**Washington** Agricultural reuse (food crops): BOD: 30 mg L^{-1} 23 total coliform / 100 ml (max)	**WHO Guidelines** Unrestricted irrigation E. coli per 100 ml Root crops 1000 Leaf crops 10 000 Drip irrigation, high growing crops 100 000 1 helminth eggs per litre

(Continued)

Table 4.6 (Continued)

Water Use	Different	Guideline	Providers				
No straight contact with food or food related activities involving public	pH 6–9 (test weekly) 30 mg L^{-1} BOD (test weekly) 30 mg L^{-1} total suspended solids (test daily) 200 faecal coliform per 100 mL (test daily) 1 mg L–1 Chlorine residual minimum	–	Class B: food consumed raw without skin E. coli – 100 colony forming units per 100 ml Test once a week BOD – 25 mg L^{-1} Test once a month	Category IV: Does not require testing	No limits for fodder, fibre and seed crops or where irrigation is straight into the ground	Industrial reuse: 240 faecal coliform/ 100 ml (max)	–
Agricultural and urban use	pH 6–9 (test weekly) 10 mg L^{-1} Biochemical oxygen demand (BOD) (test weekly) 2 Turbidity units (NTU) (continuous testing) No detectable faecal coliform per 100 mL (test daily)	E. coli <10 cfu per 100 mL Monitored monthly Turbidity <5 NTU pH 6.5–8.5 Disinfection Chlorine: 0.2–2.0 mg L^{-1} residual	TSS – 35 mg L^{-1} Test once a month Turbidity – Do not Test Class C: food consumed raw without skin watered straight into the ground E. coli – 1000 colony forming	Physical/Chemical suspended solids 20 mg L^{-1} (150 mg L^{-1} for stabilization ponds) – weekly Category III: Intestinal nematode 1 eggs per litre – monthly Faecal Coliforms or E. coli No test required –	Pasture, cleaning roads, non-food crops: Coliform limits: 23 per100 mL 240 per 100 mL in more than one sample in any 30-day period	Restricted urban reuse: BOD: 30 mg L^{-1} 240 total coliform per 100 ml (max)	–

	1 mg L^{-1} Chlorine residual minimum (continuous testing)		units per 100 ml Test once a week BOD – 25 mg L^{-1} Test once a month TSS – 35 mg L–1 Test once a month Turbidity – Do not test	twice monthly Physical/Chemical suspended solids 35 mg L^{-1} (150 mg L–1 for stabilization ponds) – monthly			
Indirect contact with food and restricted public access	pH – 6–9 test weekly BOD – 10 mg L^{-1} test weekly Turbidity – 2 NTU test continuous Coliform – No detectable test daily	–	Class D: non-food crops: E. coli -10 000 Colony forming units per 100 ml Test once a week BOD – 25 mg L^{-1} Do not test TSS – 35 mg L^{-1} Do not test Turbidity – Do not test	Category II: Intestinal nematode 0.1 eggs per litre – fortnightly Faecal Coliforms or E. coli 1000 colony forming units per 100 ml – weekly	Food crops (no contact with edible portion) fish hatcheries, restricted recreational use: coliform limits: 2.2 per 100 mL 23/100 mL in more	–	Restricted irrigation Labour-intensive, high-contact agriculture E. coli per 100 ml 10 000 Highly mechanized agriculture 100 000

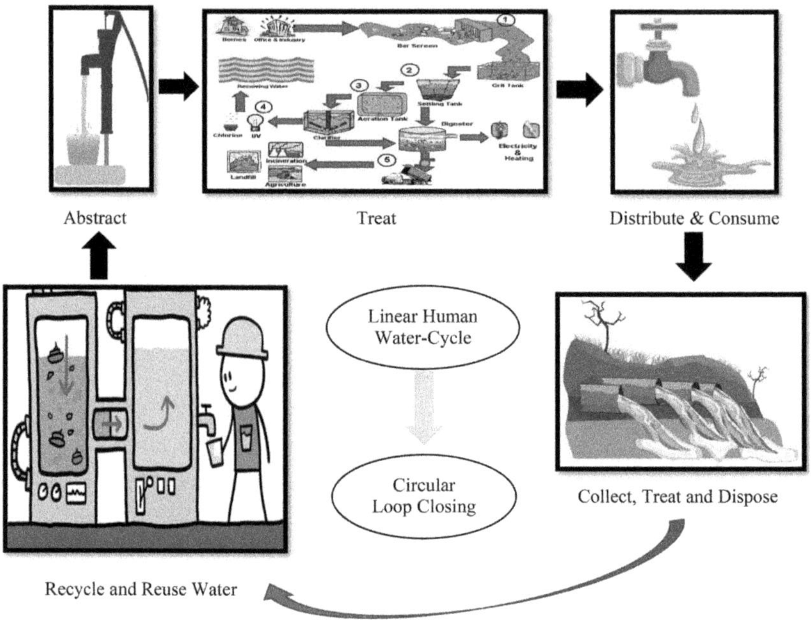

Figure 4.2 Illustration of linear water cycle to circular water cycle [1].

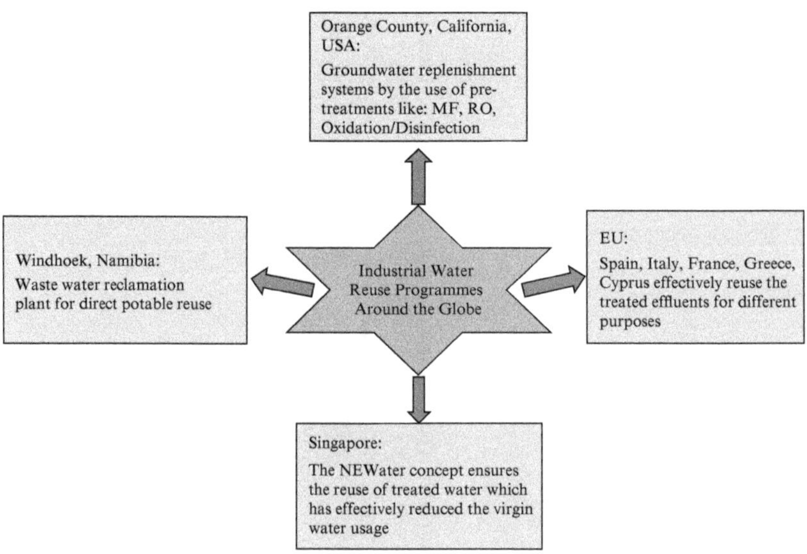

Figure 4.3 Examples of programmes for industrial wastewater reuse globally [1].

4.4 Challenges

The circular economy concept coupled with sustainability approaches have created a better connection between markets and customers. It not only provides a better business model but also reduces pressure on the natural resources ensuring their long life. Shifting from the conventional linear economic model to circular ones is a more feasible option for the stakeholders [1]. Formulation of new rules, policies and regulations will support the implementation of these concepts on large scales worldwide. The incentives provided under these policies will act as strong motivators and will be able to bring rapid changes in the perspectives of people [1]. The hurdles in collaborated application of all possible solutions can be listed as; lack of technological advancements, orthodox opinions of the people, economic considerations, safety in application of new processes and implementation of strict regulations [1]. These challenges emerge because of unclear responsibilities of the stakeholders and undefined procedures to enforce water reuse [3].

4.4.1 Technical Aspects

Industries all around the world have diversified approaches in handling the waste they produce. Few industries are able to recycle and reuse the water continuously reaching the zero net water consumption status, while few struggle to touch zero pollution [1]. Few industries have efficient manufacturing supported by reusing treated wastewater. For example, textile industries are found to use the concept of field-to-fashion largely. On the other hand, small to medium enterprises (SME) are still not able to find ways to deal with the waste they produce [1]. When the treated wastewater is supposed to be used for agriculture, the barriers are to find out which crops can be irrigated with what quality of treated wastewater, how will the soil get impacted with such practices, etc.? But if proper technologies are applied, the soil can become rich in nutrients like nitrogen, phosphorus and potassium which will reduce the cost of fertilizers [3]. The key challenge to use the treated wastewater is to avoid the access concentration of pathogens, toxic chemicals, bioaccumulation, etc. [3].

4.4.2 Economic Aspects

Circular economy theory, when applied to reuse treated industrial wastewater, offers a complete deal. Nutrient and energy recovery balances the cost of treating the substrate and also generates enough profit for the company. The positive environmental effects attract more users and maintains a feeling of harmony in the public and the surroundings [1]. The ultimate task is to ensure that the treatment and reuse of water resources is cheaper than using a virgin source. If using a resource is less costly it will eventually promote a throw-away society, the "cradle to grave" cycle of water will be unstoppable. To guarantee a "cradle to cradle" cycle for treated industrial wastewater, the overall treatment process should be affordable and the final product must find a use in some other sector [1]. The water related problems for a region may vary depending on the geographic location, climatic changes, economic status, etc. Hence, generalization of the overall cost of treating and reusing the industrial wastewater is not correct. To formulate a water reuse programme, several factors should be considered. These factors are; infrastructure and technological availability, the influent water composition of the industry, treatment procedures adopted, energy needs to operate the plant, operational and permit costs, customer demands, etc. [1]. The reason behind less water reuse projects worldwide is that there has been fewer studies evaluating the socio-economic factor of carrying out such projects. If the social advantages and benefits lingering with it are calculated in terms of the money it can potentially generate, there will more interested associations taking up such projects [8].

4.4.3 Social Aspects

The involvement of the public in implementing any water reuse scheme from industry is a critical factor. Public perception about such water vary diversely, so winning the trust of the locals is essential [9]. Recent environmental changes have made most people more aware of the impacts caused by the use of various utilities. So, reusing an already used resource is accepted as a better alternative which is more eco-friendly than construction of dams, desalination processes, etc. [1]. Government has a very important role to assure the public that the water being supplied for use is safe. People's trust is necessary to keep them motivated to use the treated water from industry. The key activities to evolve the social perspective are given below [1]:

a) Collection and management of all the necessary information related to the final treated wastewater produced and provided for different use.
b) Secure trust of the people by strong commitments from the authorities and also ensure to keep them enthusiastic about reusing the same water.
c) Involvement of different organizations to encourage public involvement of opinions.
d) Establish a firm and impartial decision making framework.
e) Promote more public activities to initiate further research and development in this area of study.

Often it is found that it is not the public who hesitate to reuse treated water again, it is the notion of the authorities that they will fail to pursuade the public to use such treated water [1]. Adapting to the use of treated industrial wastewater will create more jobs, spread positive changes in society and the environment and will also contribute in reaching the sustainability Goals [1].

4.4.4 Risk Factor

There has been less work on the subject of risk evaluation in using treated industrial wastewater. Due to the presence of many unknown factors, concerns are developing in using such resources [1]. Treated industrial wastewater often has a varied composition involving heavy metals, human and animal pathogens, pharmaceutical residues, access nutrient and sodium concentrations, etc. For this reason, before application, e.g., in the agricultural sector it must be evaluated with proper care [1]. Although the use of treated wastewater from industry is a feasible option economically, its effect on the market price of the produce needs to be evaluated [1]. To reduce the risk factor there has to be the creation of a solid systemic framework. The framework should be reinforced to tackle any accidental barriers appearing in the way. The system must be [1]: (a) resilient (capable of recovering quickly, immediately after facing a systemic setback), (b) redundant (there has to be incorporation of backup or extra process systems to ensure continuous working of the plant even if one system is down), (c) robust systems (incorporation of available technologies which can tackle collective pollutants and failures if any).

4.4.5 Policy Frameworks

Water is a free resource, it is a gift of nature to humanity. But its increased usage by all has made it necessary to charge for its use. Although there are different kinds of abstractions, i.e., surface water (rivers, streams, ponds, etc.), ground water (pumps, bore-wells etc.), they are often charged in a similar way. Recent disasters in the world created a need for proper rules and regulations to deal with the environmental issues [1]. The legislations formed to date have agreed to regard sustainability standards only if safety is not comprised. Hazard analysis critical control point schemes need to be clearly specified to ensure the safety of the people handling treated wastewater [8]. Formulating regulations will return the maximum volume of water from the industry in potable form. Industries are often found to take more care to ensure best quality standards of treated water when they use conventional sources. In spite of higher levels of contamination, these frameworks would maintain the enthusiasm for the industries to recycle more residual water than using fresh sources [1]. A "polluters pay" principal can play a key role [2]. The policies can involve fining the polluters, imposing tariffs based on the quality of waste generated, etc. This will not only motivate people to be cautious but will also create a positive impact on the natural resources and surroundings [1]. The different incentives can be, e.g., availability of funds to promote more research, different subsidies to promote increased recycling and reusing and, if the required standards are not reached, then direct closing of the plants to be carried out [2]. To promote and keep the institutions enthusiastic about the treatment and reuse work, governmental authorities can grant them certification for a limited time only, and to renew them if the certificate expires [9]. The circular economy concept can be used as an important tool to design these laws [1]. One of the most important keystones in the creation of any policy or scheme is the composition of influent water, the required quantity of treated water for specific purposes and unwavering dependency on the operations used [1]. Policies should also pave the way to develop better infrastructures for the implementation of reuse schemes in industries. Lack of infrastructure often forces industries to adopt disposal of wastewater produced rather than reusing it [8].

4.4.6 Possible Solutions

Industries should be located in such a way that they are saturated in a particular area. In that way, one industry could use the resources coming out of another industry. Proper research and planning must be undertaken in order to ensure the closing of the water cycle amongst these industries. Proper permits and monitoring systems are required to maintain checks on such industries. This circular way of operation will not only help the industries to get rid of the waste they produce after providing minimal treatment but will also serve them economically when they sell their effluents to other industries [2]. The environmental impact will be immensely positive. The human and animal life will flourish ensuing the sustainability of all kinds of resources and healthy living conditions for all [2]. Continuous awareness campaigns must be carried out amongst the public, to not only promote preventing water waste but also initiate the spark to reuse the same water in a variety of ways [2]. These programmes will motivate the industries to adopt cleaner production practices and will also build a sustainable network of water [2]. There can be the adoption of agent-based modelling systems to help people understand the advantages of reusing treated wastewater [8]. The concept of industrial wastewater management with a focus on reuse (IW2MC→R), is a way to utilize the wastewater coming from industries into toilet flushing, gardening and cleaner production practices [10]. The state-of-the-art theory; multi-criteria multi-decision-makers framework integrating multi-criteria decision analysis (MCDA) backed by game theory presents a holistic way to formulate water reuse schemes. This theory can be implemented in any industry and a greater number of factors can be analysed together to come up with a feasible programme for that particular industry [8].

The figure given below illustrates the interconnection of three driving pillars considered in game theory. Critical analysis of all the factors together generates a public favoured and environment friendly scheme of water reuse programme [8, 11].

Often, the perceived fears within the public regarding the use of such treated wastewater from industry are stronger than their belief in the process efficiencies. And hence constant efforts are to made to ensure that the public is well informed

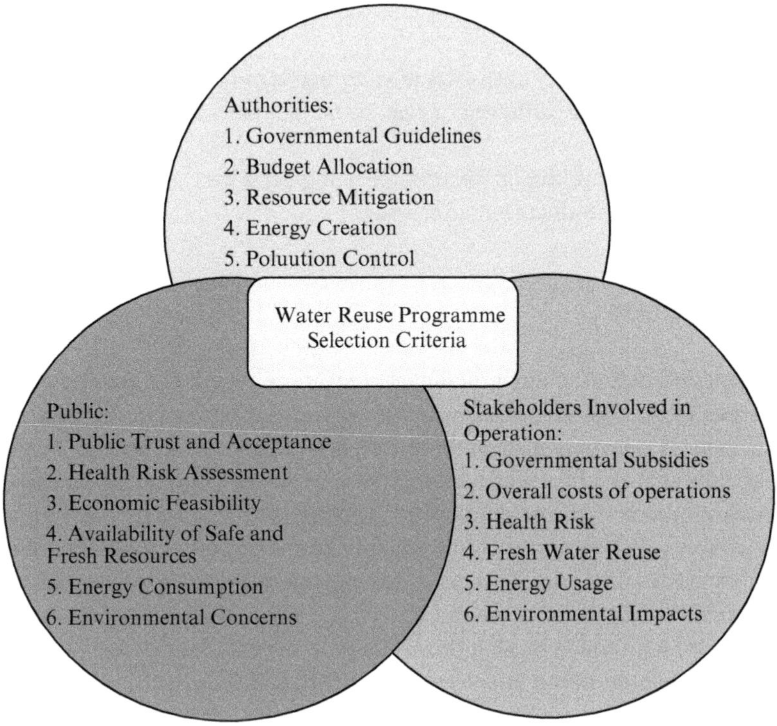

Figure 4.4 Illustration of game theory used in deciding the best water reuse scheme [8].

about the water quality being provided to them for potable use. Trust in authorities is a game changer to satisfy this factor [9]. Another method to come up with fruitful schemes can be the decision support system (DSS). The factors that are taken into account while applying this method are discussed below [5]:

1) Wastewater characteristics; depends on the origin: urban, industrial, agriculture, mixtures and also on the sewage network; the concentration of pollutants and its variation in the composition are the deciding factors for what treatment to be applied.
2) Amount of treated water required; depends on the requirement of the user, budget of the buyer, water laws barriers.
3) Present policies and framework; standards mentioned by different authorities like EU, WHO etc. as discussed in the present chapter.
4) Technological availability; this factor also has an economic side to it.
5) Social aspect; the reaction of the public to different policies holds huge importance.

6) Centralization and decentralization of treatment plants according to the plan of the city is also a crucial factor.
7) Influent and effluent quality of the wastewater needs to be analysed perfectly.
8) Nutrient integration for different usage must be assessed for industries separately.
9) Scope of reuse and recycling of wastewater being produced and measures to manage unusable by-products left after reuse.

4.5 Future Scope

Before the implementation of any scheme or programme, there is a requirement for overall study of the region including all the factors as discussed in the present chapter. As water is unevenly distributed throughout the world forming a single policy generically will not be a feasible idea. So, any city or country has to carry out their own research with the support of available studies to deal with their unique water issues [1]. Every year the expansion plan of any city should incorporate the treated wastewater reuse from all industries operating in the nearby areas. This is only possible if all social, economic and environmental aspects together are taken into account with the help of all the available sources of knowledge [3]. The research to be carried out in the future must be directed as follows [3]:

a) The present and future obtainability of the treated wastewater from region specific industries must be assessed.
b) Impact assessment of prolonged use of a particular treated wastewater for a specific use must be carried out.
c) Continuous assessment of socio-economic status must be carried out to analyse the profits involved.

4.6 Conclusion

Industrial revolution brought an increased global use of water. Awareness related to the environmental issues resulted in the creation of strict laws by different nations. These laws have forced the industrial sector to cut down their water usage and adapt to reusing the same water after treatment [2]. The recycling and reuse of industrial wastewater has tremendous potential to curb the water crisis around the world. The incorporation of a circular economy and sustainability concept in this sector will benefit all the three pillars, i.e., social, economic, and environment, of the society [1]. The challenges in the implementation of the policies discussed in this current chapter can be removed with further case-by-case based studies. There has to be a holistic approach backed by all developed and developing countries towards this new sector of opportunities [1]. However, there needs to be a check for the points mentioned below to come up with a good treated industrial water usage strategy [5]: technological considerations, political subsidies and availability of funds from governmental/non-governmental authorities, design of all unit operations involved, policies enforced to reach the needed standards, thorough studies of the present and future impacts of adopting any new technique, availability of infrastructure, management and commissioning authorities, disposal of unused components to be prefixed. Extending responsibilities with public and private sector partnership (PPP) in closing the water loop will be beneficial immediately [3, 6].

References

1 Voulvoulis, N. (2018). Water reuse from a circular economy perspective and potential risks from an unregulated approach Elsevier B.V 2468–5844. doi: 10.1016/j.coesh.2018.01.005.

2 Vishvanathan Takashi Asano, C. (2001). The potential for industrial wastewater reuse encyclopedia of life support systems. *(EOLSS) WASTEWATER RECYCLE, REUSE, AND RECLAMATION* I: https://www.eolss.net/sample-chapters/c07/E2-14-02-04.pdf.

3 Elbana, T.A., Bakr, N., and Elbana, M. (2017). *Reuse of Treated Wastewater in Egypt: Challenges and Opportunities*. Springer International Publishing AG. doi: 10.1007/698_2017_46. https://www.researchgate.net/publication/317289855_Reuse_of_Treated_Wastewater_in_Egypt_Challenges_and_Opportunities.

4 Riggio, V. A., Ruffino, B., Campo, G., Comino, E., Comoglio, C., & Zanetti, M. (2018). Constructed wetlands for the reuse of industrial wastewater: A case-study. *Journal of cleaner production*, 171, 723–732.

5 Salgot, M., & Folch, M. (2018). Wastewater treatment and water reuse. *Current Opinion in Environmental Science & Health*, 2, 64–74.

6 Maryam, B., & Büyükgüngör, H. (2019). Wastewater reclamation and reuse trends in Turkey: Opportunities and challenges. *Journal of Water Process Engineering*, 30, 100501.

7 Zajda, M., & Aleksander-Kwaterczak, U. (2019). Wastewater treatment methods for effluents from the confectionery industry–an overview. *Journal of Ecological Engineering*, 20(9), 293–304.

8 Ghernaout, D., Elboughdiri, N., and Al Arni, S. (2019). Water reuse (WR): dares, restrictions, and trends. *Applied Engineering* 3 (2): 159–170. doi: 10.11648/j.ae.20190302.23 http://www.sciencepublishinggroup.com/j/ae.

9 Saliba, R., Callieris, R., D'Agostino, D. et al. (2018). Stakeholders' attitude towards the reuse of treated wastewater for irrigation in Mediterranean agriculture. *Agricultural Water Management* 204 (2018): 60–68 doi: 10.1016/j.agwat.2018.03.036.

10 Bauer, S., Dell, A., Behnisch, J. et al. (2020). Water-reuse concepts for industrial parks in water stressed regions in South East Asia. IWA Publishing 2020. *Water Supply* 20 (1). http://iwaponline.com/ws/article-pdf/20/1/296/647108/ws020010296.

11 Ghernaout, D., Elboughdiri, N., and Ghareba, S. (2019). Drinking water reuse: one-step closer to overpassing the "Yuck Factor". *Open Access Library Journal* 6: e5895. 10.4236/oalib.1105895.

5

Biological and Microbiological Characteristics of Activated Sewage Sludge

Vipin Singh[1,2] *and Satarupa Dey*[3,*]

[1] *Environmental Science and Engineering Department, Indian Institute of Technology Bombay, Mumbai, India*
[2] *Global Product Compliance, India*
[3] *Department of Botany, Shyampur Siddheswari Mahavidyalaya, Howrah, West Bengal, India*
[*] *Corresponding author*

5.1 Introduction

The activated sludge technique is currently the most popular method for biological treatment of wastewater. Proposed by Arden and Lockett in 1913 this process has undergone several changes and developments, and is widely adopted by countries around the world. The process of activated sludge is broadly divided into two separate phases which are the aeration phase and sludge settlement phase. In the aeration phase no settlement is allowed while in the settlement phase it is allowed followed by removal of the sludge. The wastewater from the primary settlement tank is passed into the aerobic tank containing a wide range of microbial population. The tank is aerated by surface agitation or via diffusers to add oxygen which is essential for the growth of aerobic microorganisms in the reactor. This aeration is also vital for the maintenance of the microbial flocs and maximizes the contact time between the surface of the floc and the wastewater. Oxygen also facilitates mass transfer and helps to disperse the metabolic products trapped within the flocs. In the second stage the flocculated biomass settles to form sludge which clarifies the effluent from solids which is discharged as final effluent. In an activated sludge process around 0.5 and 0.8 kg dry weight of sludge is produced for every kilogram of biological oxygen demand (BOD) removed. Most of the activated sludge is then returned so that there is always an adequate microbial population to oxidize the upcoming wastewater.

Application of Sewage Sludge in Industrial Wastewater Treatment, First Edition. Edited by Maulin P. Shah.

Microbial floc formation is crucial in any activated sludge process as they help in adsorption of the organic matter in the wastewater and also help in rapid and effective separation of effluent in the sedimentation tank itself. It is noted that any changes in the operation parameters may lead to changes in the nature of the microbial floc which leads to poor settlement resulting in turbid effluents and subsequent loss in biomass. The sludge generated in the process is entirely composed of microbial biomass and adsorbed particulate matter. The main component of activated sludge consists of reactor, the activated sludge which comprises of flocculant suspension bacteria and other microfauna and microflora. Apart from this an aeration and mixing system and a sedimentation tank is required where the final settlement of activated sludge (AS) occurs and separates microbial biomass. Some of the activated sludge is recycled back to the reactor to maintain the microbial population. Activated sludge works in food limited conditions and each microbe uses its own cellular content and in turn reduces the biomass produced. However, normally, the growth of the microbial population and the accumulation of non-biodegradable solids result in an increase in the amount of AS produced.

The main two principles for the removal mechanism in an activated sludge process is assimilation and mineralization. The assimilation process is carried out by utilizing the waste materials to create a biomass. The colloidal and soluble part of the wastewater is transformed into biomass, which is settled out. Mineralization is associated with the process of degradation of waste to inert end products which are left in solution in the effluent and require longer aeration times. The assimilation process leads to rapid removal of BOD associated with high production of sludge thus the treatment cost is high.

The main characteristics are represented in Table 5.1 [1]

Table 5.1 Main characteristics of activated sludge.

	Activated sludge characteristics
Capital cost	Less
Operating cost	Costly
Area of land	Requires less area to operate so the process is advantageous where land availability is restricted or expensive
Influence of weather	Hot and humid weather works perfectly; effectiveness reduces in dry weather; less affected at low temperatures
Technical control	High; the microbial activity can be closely controlled; requires skilled and continuous operation
Nature of wastewater	Sensitive to toxic shocks and changes in loading; trade wastewaters can lead to bulking problems
Nuisance	Low smell with reduced fly problems; noise is a problem
Final effluent quality	Poor nitrification but low in suspended solids

This present chapter mainly deals with the biological properties of the activated sludges and discusses the microbial communities present in such sludges. The seasonal variation of the microbial communities in activated sludge is also discussed in detail. Along with the abundance of different microbial groups, their role and physiological activities in activated sewage sludge are also reviewed.

5.2 Biological Properties of Activated Sludge

Carbohydrates, proteins, and lipids are the chief constituents in municipal wastewater which provide nutrients to the bacterial community and helps in floc formation. In several researches it was found that carbohydrate was associated with large particles of size > 63 μm. Huang and Gu [2] also observed that size fractions smaller than 0.1 μm contribute nearly 62% of the TOC which is mainly associated proteins or carbohydrates.

Nitrogen can be present in both inorganic form that is in form of ammonium or nitrate or else present in organic forms. Generally, nitrates present in activated sludge are in soluble form which is of the most concern for groundwater pollution. On the other hand, inorganic nitrogen in the form of ammonium is volatile and is lost. Organic nitrogen found in the activated sludges can be considered as low release form and needs to degraded by microorganisms, or mineralized to inorganic NH_4 and NO_3. Some other sludge constituents, including phosphorus (P), calcium (Ca), magnesium (Mg), and iron (Fe), are known to form insoluble compounds with sludge solids, and are present at high concentration. Other sewage sludge constituents, such as potassium and sodium, being water-soluble and are normally discharged with the treated wastewater. Compounds such ashumic acids, surfactants, amino acids, tannic acids, volatile fatty acids (VFAs), DNA, and RNA were found in excess.

5.3 Microbial Community Structure of Activated Sludge

The studies on activated sludge involves an inventory of diversity of microbial communities and is considered to be functionally an important population. The microbial community of activated sludge was done both by culture dependent and culture independent methods. The culture dependent methods however were easier, but it does not give a thorough idea of the microbial community of activated sludge due to inability to grow the majority of microorganisms in any specific culture conditions. With the advent of different molecular biology methods such as polymerase chain reaction (PCR) techniques the study of microbial diversity has been revolutionized.

PCR based techniques provide us with detailed information on the presence, diversity, and expression of ribosomal and protein-coding genes. The advent of the "–*omics* era" was considered to be a turning point in the study of both phylogenetic and functional diversity of activated sludge systems which has given us an unprecedented access to genes and genomes and helps us in understanding the diversity of the microbial community. The advent of high throughput sequencing (HTS) techniques such as 454-pyrosequencing and Illumina results in the production of millions of sequence reads in a cost-effective way. Applications of metagenomics and metatranscriptomics approaches helps in massive DNA sequencing of all genes present in a microbial community and their genomic potentials have been unrevealed. The most widely used methods such as clone library, DNA fingerprinting, fluorescent in situ hybridization, and quantitative polymerase chain reaction (qPCR) studies based on functional genes or 16S rRNA gene segments have helped in having an idea on the microbial community of activated sludge.

The activated sludge process is the most common method which has been applied in biological process of any wastewater treatment plants (WTP). This activated sludge floc consists of highly complex microbial communities comprising eukaryotes, bacteria, archaea, and viruses. These activated sludges are dominated by bacteria which play a crucial role in the removal of organic pollutants and nutrients including nitrogen and phosphorus. These bacterial communities can tolerate adverse environmental impact, toxicity, and oxygen depletion. They also perform a crucial role in biological nitrification and oxidizes ammonia to nitrate and nitrite then to nitrogen via denitrification.

The process of nitrification is primarily done by two separate groups of microorganisms: (i) ammonia-oxidizing microorganisms [3], and (ii) nitrite-oxidizing bacteria [4]. Several studies have been conducted on ammonia-oxidizing microorganism, nitrite-oxidizing bacteria, and denitrifying bacteria. They contain several biomarker genes such as ammonia monooxygenase (amo) [5], nitrite reductase subunits (nirK and nirS) [6].

Jin et al. [7] reported the presence of four bacterial phyla-*Proteobacteria*, *Actinobacteria*, *Bacteroidetes* and *Firmicutes* along with the presence of *Actinobacteria*, *Acidobacteria* [8] and *Verrucomicrobia*. Other dominant phyla include *Chloroflexi*, *Planctomycetes*, *Verrucomicrobia*, and *Actinobacteria* [9, 10].

Yu and Zhang [11] in their study suggested that bacteria were dominant accounting for nearly 92% and 69% of DNA and cDNA sequences respectively. Whereas, eukaryote accounts for approximately 43% and 30.97% of total sequences in DNA and cDNA, respectively. The bacterial community was dominated by *Proteobacteria*, *Actinobacteria*, *Bacteroidetes*, and *Firmicutes* representing nearly 22%, 15%, 6%, and 3% of SSU rDNA reads respectively. Both *Verrucomicrobia* and *Nitrospirae*, showed high abundance in protein coding DNA reads. Among Archaea *Euryarchaeota* also had a very high percentage in SSU rDNA (19.38%). *Actinobacteria*, *Firmicutes*, *Planctomycetes* and *Euryarchaeota* had % SSU rRNA/% SSU rDNA less than one. However, the same ratio was higher than one in the cases of *Proteobacteria*,

Acidobacteria, *Bacteroidetes*, *Verrucomicrobia*, *Chytridiomycota*, *Ascomycota*, and *Basidiomycota*. *Proteobacteria* accounts for nearly 37.5% and 56.1% of annotated bacterial SSU rDNA and rRNA sequences, which was followed by phyla such as *Bacteroidetes*, *Verrucomicrobia*, and *Actinobacteria*.

The microbial community was dominated with *Nitrosomonas, Nitrosospira, Methylocystis* and *Methylosinus* having high activity of ammonia monooxygenase. Presence of *Nitrosomonas*, followed by *Nitrosospira, Methylocystis*, and *Anaeromyxobacter* accounts for the nitrification enzyme. *Acidovorax, Cupriavidus, Leptothrix, Alicycliphilus, Paracoccus* and *Escherichia* reported hydroxylamine reductase activity. On the other hand, *Riemerella, Dyadobacter, Dechloromonas, Candidatusaccumulibacter* and *Acidovorax*reported high nitrous oxide reductase activity.

The microbes of activated sludge include ammonia oxidizing bacteria (AOB) [12, 13], nitrite oxidizing bacteria (NOB) [14], denitryfiers [13, 15], phosphorus accumulating organisms (PAOs) [16] and filamentous bacteria [17]. The wastewater treatment plants contain *Curvibacter, Azoarcus, Thauera, Zoogloea*, and *Accumulibacter* which are mainly denitrifiers, *Tetrasphaera* and *Accumulibacter* which are reported to be phosphorus accumulating organisms and filamentous bacteria such as *Microthrix parvicella* and *Gordonia*. Low-abundance of antibiotic resistant bacteria and potential pathogens are also environmentally important populations which have also received increasing attention.

Zielinska et al. [15] identified the presence of 38 orders from microbial consortia of WTPs which include *Anaerolineales*, *Rhodocyclales*, *Burkholderiales*, *Rhizobiales*, *Planctomycetales*, etc., and six core genera such as *Ferruginibacter*, *Prosthecobacter*, and *Zooglea*. The presence of denitrifying populations such as *Azoarcus, Thauera, Curvibacter*, and *Dechloromonas* was also evident [18, 19]. *Candidatus accumulibacter* belonging to the family Rhodocyclaceae were designated as phosphorus accumulating organisms and were also identified along with *Pseudomonas* as having denitrifying properties. *Halomonas* was also present in large numbers comprising 5% of the microbes present in activated sludge, Nguyen et al. and Kristiansen et al. [20, 21] reported the presence of *Tetrasphaera* belonging to the family *Intrasporangiaceace* which was reported to contain functional genes for denitrification. Similarly, Nielsen et al. [22] reported the abundant presence of *Dechloromonas*spp. which was also a denitrifier and a putative PAO on the basis of its phenotype.

Zielinska et al. [15] also reported that 70% of the alpha diversity belonged to *Proteobacteria* in which the *Alphaproteobacteria*, and *Betaproteobacteria* were the most abundant group. In the largest group at order level within *Alphaproteobacteria* was *Rhizobiales*, while among *Betaproteobacteria* it was *Burkholderiales*. Their findings also show lesser abundance of gamma- (6.5%) and deltaproteobacterial sequences (9.9%) compared to the previous studies conducted by the previous researchers. Apart from them bacterial reads belonging to the order *Rhodobacteriales*, *Rickettsiales* and *Rhodocyclales* was also reported in this study.

The microbes belonging to *Betaproteobacteria* and *Alphaproteobacteria* were reported to contain genes coding for periplasmic nitrate reductase (napA) and gene coding for membrane bound nitrate reductase (narG) [23]. However, *Actinobacteria* accounts for nearly 11% narG. The class *Betaproteobacteria* consists of families such as the *Comamonadaceae* and *Rhodocyclaceae* families which forms a dominant group of microbial community [24].

In a much later study Zhang et al. [25] reported the presence of bacterial OTU to be assigned to 14 different phyla including *Acidobacteria*, *Chloroflexi*, *Actinobacteria*, *Bacteroidetes*, *Firmicutes*, *Planctomycetes*, *Chlorobi*, *Saccharibacteria*, *Verrucomicrobia*, and *Proteobacteria*. *Proteobacteria* consisted of a total of 47% of the OTUs, followed by *Bacteroidetes* (30%), *Firmicutes* (7%), *Acidobacteria* (2.2%) and *Chlorobi* (1.2%). Amongst *Proteobacteria* classes *Gammaproteobacteria* (25%) and *Betaproteobacteria* (24%) was most prominent. In addition, *Flavobacteriia* (18%) and *Cytophagia* (13%) were also significantly abundant. However, a previous study done by Xu et al. [26] in which it was stated *Betaproteobacteria* was the largest class. *Gammaproteobacteria* being more sensitive to various antibiotics present in the sludge are usually present in much lower quantities [27].

The details of the microbial community available in activated sludges are listed in Table 5.2.

5.4 Seasonal Variation of Microbiological Characteristic of Activated Sludge

Seasonal variation, particularly variation in temperature, is a vital factor that alters the microbial communities of activated sludge; however, the underlying mechanisms of such a shift is still a topic of research and not fully understood. Several factors like total nitrogen, HN3-H, and chemical oxygen demand (COD) were improved with the increase in temperature from 20 to 35 °C. However, at reduced temperature total nitrogen and total phosphorus removal processes were negatively impacted. Also, the extracellular polymer substance (EPS) composition and sludge index of activated sludge was altered with the change in temperature. The structural and functional potentials of the microbial community was also altered with a change in temperature. Ai et al. [30] reported a total of 147 phyla, 2162 and 13 015 genera and species respectively indicating high taxonomic diversity in activated sludge. At 20 °C the dominant genera which were reported in activated sludge included *Sphingopyxis, Bradyrhizobium, Candidatus, Saccharimonas, Mesorhizobium* and *Bosea*. Apart from them other genera such as *Sphingomonas*, *Thauera*, *Azoarcus*, *Candidatus,Contendobacter*, *Competibacter* and *Pyrinomonas* were also reported. Among the different genus reported *Sphingopyxis,Candidatus* and *Competibacter* have the capacities to secrete EPS. This EPS played a vital

Table 5.2 Microbial abundance in activated sludge.

Phylum	Class	Genera	Method of study	References
Halobacterota, Micrarchaeota, Nanoarchaeota				[28]
Proteobacteria				
Bacteroidota				
Patescibacteria, Myxococcota				
Actinobacteriota				
Planctomycetota				
Chloroflexota, Acidobacteriota				
Firmicutes				
Proteobacteria		*Nitrosomonas*	Illumina HiSeq 2500	[29]
Nitrospirae		*Hyphomicrobium*		
Chloroflexi		*Nitrospira*		
Chlamydiae, Chlorobi, Chloroflexi, Elusimicrobia, Ignavibacteriae, Latescibacteria, Parcubacteria, Spirochaetae		*Bdellovibrio*		
		Thauera		
		Halochromatium		
		Terrimonas		
Armatimonadetes, Bacteroidetes, Chlamydiae, Chlorobi, Cyanobacteria, Fibrobacteres, Verrucomicrobia		*Ferruginibacter*		
		Dechloromonas		
		Hyphomicrobium, Phaeodactylibacter		
		Parafilimonas		
		Nitrosomonas Nitrospira		

(Continued)

Table 5.2 (Continued)

Phylum	Class	Genera	Method of study	References
Proteobacteria, Actinobacteria, Acidobacteria Bacteroidetes		*Sphingopyxis, Bradyrhizobium,* *Candidatus Saccharimonas, Mesorhizobium, Bosea, Niastella, Acidovorax, Alicycloiphilus* *Sphingomonas, Thauera, Azoarcus, Candidatus Contendobacter,* *Candidatus Competibacter Pyrinomonas* *Piscicoccus, Kineosphaera, Microlunatus, Dehalobacter, Nitrospira, Tetrasphaera, Nakamurella,* *Propionicella Friedmanniella*	Illumina sequencing	[30]

Acidobacteria, Chloroflexi, Actinobacteria, Bacteroidetes, Firmicutes, Planctomycetes, Chlorobi, Saccharibacteria, Verrucomicrobia, Proteobacteria.	*Gammaproteobacteria* (25%) *Betaproteobacteria* (24%) *Flavobacteriia* (18%) *Cytophagia* (13%) *Clostridia* (18%) *Sphingobacteriia* (12%) *Epsilonproteobacteria* *Mollicutes* *Bacteroidia* *Chloroflexia* *Cytophagia* *Nitrospira* *Negativicutes* *Ignavibacteria* *Spartobacteria* *Spingobacteriia*	*Flavobacterium* sp. (13%) *Pseudomonas* sp. (8%), *Alcaligenes*, *Acinetobacter* sp. *Legionella* *Acidovorax* *Dokdonella* *Bacillus* *Lactobacillus*, *Terrimonas* *Pseudofulvimonas* *Nitrospira* sp., Nitrosococcus sp., *Nitrosomonas* sp. *Brevundimonas* sp. *Pedobacter* sp., *Chryseobacterium* sp., *Comamonas*sp., *Archrobacter*sp., *Stenotrophomonas* sp., *Dyadobacter* sp., *Caulobacter* sp., *Cohnella*sp., *Massilia*sp., and *Exiguobacterium*sp.	Illumina Miseq sequencing platform	[25]

(Continued)

Table 5.2 (Continued)

Phylum	Class	Genera	Method of study	References
Proteobacteria (12.3–58.5%), *Acidobacteria* (1.8–35.1%), *Chloroflexi* (2.8–37.7%) *Bacteroidia* *Actinobacteria* (0.7–6.8%), *Synergistetes* (0.02–5.6%)		*Caldilinea*, *Dechloromonas* *Thiobacillus*, *VadinCA02*, *Thauera* *Nitrospira*	Illumina shotgun DNA library	[31]
Proteobacteria *Nitrospirae* *Bacteroidetes* *Actinobacteria* *Firmicutes* *Euryarchaeota*	*Betaproteobacteria* (46.19) *Gammaproteobacteria* (11.14) *Alphaproteobacteria* (8.19) *Deltaproteobacteria* (1.51) *Epsilonproteobacteria* (0.07) Nitrospira (15.4) *Flavobacteriia* *Sphingobacteriia* *Cytophagia* *Bacteroidia* *Ignavibacteria* *Actinobacteria* *Gemmatimonadetes* *Acidobacteriia* *Solibacteres* *Clostridia* *Bacilli* *Negativicutes*	*Rhodobacter* *Caulibacter* *Pseudomonas* *Geobacter* *Rhodopseudomonas* *Thauera* *Dechloromonas* *Bordetella* *Nitrosomonas* *Nitrospira* *Mycobacterium* *Placntomyces* *Bradyrhizobium* *Burkholderia* *Xanthomonas*	Illumina HiSeq. 2000	[32]

Actinobacteria *Proteobacteria* *Nitrospirae* *Bacteroidetes* *Firmicutes* *Verrumicrobia* *Spirocheate* *Cyanobacteria* *Deinococcus-Thermus* *Chloroflexi*		*Nitrospira* sp., *Nitrosomonas* sp. *Pseudomonas putida* *Nitrosovibrio*	GeoChip 4.2	[33]
Actinobacteria *Proteobacteria* *Verrumicrobia Firmicutes Bacteroidetes* *Fusobacteria* *Nitrospirae* *Planctomycetes* *Deinococcus-Thermus* *Chloroflexi* *Tenericutes* *Euarchaeota*	*Deltaproteobacteria* *Epsilonproteobacteria* *Betaproteobacteria* *Gammaproteobacteria* *Alphaproteobacteria* *Dienococci* *Clostridia* *Thermomicrobia* *Methanobacteria* *Methanomicrobia*	*Acinetobacter* *AkkermansiaAcidaminococcus,* *Cloacibacterim,* *MegasphaeraPrevotella* *Streptococcus,* *Trichococcus* *Gemmatimonas*	GS Junior system (Roche)	[34]

(Continued)

Table 5.2 (Continued)

Phylum	Class	Genera	Method of study	References
Actinobacteria *Proteobacteria* *Nitrospirae* *Bacteroidetes* *Firmicutes* *Verrumicrobia* *Spirocheate* *Cyanobacteria* *Deinococcus-Thermus* *Chloroflexi* *Tenericutes* *Verrumicrobia*	*Betaproteobacteria* *Gammaproteobacteria* *Alphaproteobacteria* *Deltaproteobacteria* *Flavibacteria* *Nitrospira* *Bacteroidia* *Solibacteres* *Chloroflexi* *Clostridia* *Cytophagia* *Deinococci* *Sphingobacteria*	*Mycobacterium* *Clostridium* *Hyphomicrobium* *Tissierella* *Sphingomonas* *Desulfobacterium* *Kribbella* *Deinococcus* *Saccharopolyspora* *Myroides*	Metagenomics-rapid annotation using Subsystem Technology server (MG-RAST)	[35]

role in stabilizing of granule structure and accumulation of phosphorus in activated sludge [36], whereas the denitrification process was facilitated by *Thauera, Acidovorax* and *Azoarcus* [8, 37]. When the temperature was maintained at 20 °C for a very long time new genera were dominated by genus like *Piscicoccus, Kineosphaera, Microlunatus, Dehalobacter, Nitrospira, Tetrasphaera, Nakamurella, Propionicella* and *Friedmanniella*. All these isolates were known to perform various types of function such as organic removal, biological phosphorus removal, and nitrite oxidation [38, 39]. With an increase in temperature to 35 °C the occurrence of the bacterial genus was greatly changed and *Candidatus, Contendobacter, Candidatus,Competibacter* and *Pyrinomonas* was also found, at 35 °C. The dominant genera were *Sorangium, Chondromyces, Anaeromyxobacter, Pyrinomonas, Labilithrix,* and *Rubrivivax*. As the temperature increased the density, viscosity and thickness of the film decreased which increased ion diffusion and accelerated the adsorption process. With the increase in diversity of the microbial community the metabolic process also increased which increased COD removal. However, the functional role of most of the dominant genera is not clear. As the temperature was reduced to 5 °C the microbial community shifted showing more occurance of genera like *Dechloromonas, Gemmata, Hassallia, CandidatusAccumulibacter, Opitutus,* and *Pedospaera*. Genera like *Opitutus* and *Dechloromonas* were involved in nitrogen cycling and biological phosphorus removal respectively [40, 41].

5.5 Physiological Function and Enzyme Activity of Microbiome in Activated Sludge

A wide range of enzymatic activity was seen by the microbial community seen in the wastewater. Most of the microbes perform a vital role in ammonification, nitrite and nitrate oxidation and denitrification. The process of nitrification in wastewater is carried out by two diverse groups of microorganisms. The first group is dominated by an ammonia-oxidizing microorganism [3, 42] which oxidize ammonia to nitrate and the second group is nitrite-oxidizing bacteria which further oxidize nitrite to nitrate [4]. Complete ammonia oxidizing bacteria (comammox) is also known to occur which is capable of oxidizing ammonia via nitrite to nitrate directly [43]. However, its role in wastewater treatment needs further research. Zielinska et al. [15] reported a high abundance of denitrifying bacteria which included *Curvibacter* within Comamonadaceae, apart from which genera like *Azoarcus, Thauera, Dechloromonas,* and *Accumulibacter*, were dominant in activated sludge. Under anaerobic conditions these denitrifying bacteria reduce nitrite to gaseous forms such as nitrous oxide and dinitrogen gas which in turn may reduce the wastewater nitrogen concentration.

Both DNA and cDNA show the presence of a wide range of ammonia assimilation, nitrite/nitrate ammonification, denitrification, and nitrogen fixation related genes. A wide range of enzymes such as ammonia monooxygenase, hydroxylamine reductase, hydroxylamine oxidase, nitrate reductase, nitrilase, formamidase, carbamate kinase, nitric oxide reductase, nitrite reductase, nitrous oxide reductase, and nitrogenase related DNA sequences were obtained. The genes related to ammonification amoCAB encoding ammonia monooxygenase was comparatively increased when the temperature was raised from 20 °C to 35 °C and the enzymes related with denitrification were reduced. With the decrease in temperature from 20 °C to 5 °C genes related to nitrogen metabolism was raised by a greater amount. Abundance of genes related to carbamate kinase, glutamate dehydrogenase, and glutamine synthetase was increased. Enzyme nitrite reductase (nrfA) which is associated with reduction of nitrite to ammonia along with hydroxylamine reductase (hcp) related to the reduction of hydroxylamine to ammonia was also increased.

Yu and Zhang [11] also calculated the abundance of DNA sequences in activated sludge. They reported the abundance of ammoni monooxygenase (amo), hydroxylamine reductase (har), hydroxylamine oxidase (hao), nitrate reductase (nar), nitric oxide reductase (nor), nitrite reductase (nir), nitrous oxide reductase (nos), and nitrogenase (nif). In a 2.4 Gbp DNA nir gene was found in abundance, followed by nor and nos coding gene sequence. Around 2855 sequences were annotated with nif coding genes. However, the abundance of nitrification enzymes coding gene sequences was the lowest including amo and hao. Nitrifying virus was expressed in higher amount than that of denitrification enzymes. In the case of Hydroxylamine oxidase the cDNA/DNA ratio was around 0.09. Nitrification enzyme such as amo was much higher expression activities in activated sludges which was mainly due to higher concentration of ammonia in sewage.

Xia et al. [33] reported that a total of 528 genes which were associated with phosphorus utilization were present including *ppk* genes (37.3%), *ppx* genes (57.6%) and *phytase* genes (5.1%). Exopolyphosphatase (ppx) is a highly processive enzyme catalyzing the anaerobic hydrolysis process of terminal residues of long-chain polyP to phosphate (Pi). Apart from that the genes related to a wide number of functions like carbon, phosphorus, and sulpher cycling, and also organic pollutant remediation were reported. The genes related to processes like denitrification, ammonification, nitrogen fixation, assimilatory and dissimilatory nitrogen reduction were also seen.

5.6 Antibiotic Resistant Gene in Activated Sludge

Wastewater treatment plants contain microbes from both human sources and environmental sources and can be a rich source of antibiotic resistant genes, moreover wastewater treatment plants receive water from households and pharmaceutical industries which exercise a selective pressure on antibiotic resistant bacteria and antibiotic resistant genes prior to their release into the environment and thus cause extensive pollutions [44, 45, 46].

Antibiotic resistance genes were found to be abundant in activated sludges which are responsible for the subsequent spread of various types of antibiotic resistant genes (ARG). Mobile genetic elements such as plasmid, transposons, and integrons contribute largely to the dissemination of antibiotic resistance genes in activated sludge. However, until now very few studies have been conducted on the host cells which harbours such antibiotic resistant genes. As much as 30 ARG encoding resistance to tetracycline, sulphonamides, quinolones, or macrolides were identified in activated sludge of two wastewater treatment plants of China by Mao et al. [47]. A significant enrichment of 10 ARG including tetB, tetG, tetH, tetS, tetT, tetX, sulI, sulII, qnrB, and ermC was also noticed by the same group.

In a recent study by Liu et al. [48] they identified around 22 bacterial phyla which can act as a putative host for these genes. Genus such as *Mycobacterium* and family Burkholderiaceae harbours around 14 to 50 antibiotic resistant genes. Metatranscriptome analysis showed nearly 65.8% of the identified ARGs were being expressed, highlighting that these antibiotic resistant genes were transcriptionally active. Most of the antibiotic genes were plasmid associated and displayed almost two-fold increase compared to those ARGs found exclusively within bacterial chromosomes. Several antibiotic genes associated with antibiotics such as sulfonamide, beta-lactam, tetracycline, and vancomycin have been reported by different groups of scientists [49, 50, 51]. Metagenomics analysis reported to be the most efficient method for the analysis of antibiotic resistance gene in the activated sludge [52, 53]. Liu et al. [48] in an extensive study on antibiotic resistance genes in the activated sludge reported 24 different classes of antibiotic in the activated sludge and genes associated with antibiotics like aminoglycosides, acriflavines, bacitracin, betalactams, daunorubicin, multidrug resistance (MDR), macrolide-lincosamide-streptogramin (MLS), polymyxin, and sulfonamide were also found in activated sludge. Abundance to different antibiotics classes is usually known to vary significantly. In the case of multidrug resistance, it was nearly 7.06×10^{-1}, whereas for nitroimidazole was 1.20×10^{-4}. In activated sludge multidrug resistance genes were most abundant followed by betalactams, macrolide-lincosamide-streptogramin, and bacitracin. Similar research [31, 54] reported aminoglycosides and tetracycline to be most prominent in activated sludge. There exist 20 different antibiotic resistance genes that accounted for nearly 70% of the total types of ARGs which included acriflavine (acrB, acrF), bacitracin (bacA,bceA), beta-lactam (pbp2), bleomycin (ble), daunorubicin (drrA), fosmidomycin (rosA), kasugamycin (ksgA), MDR (mdtB, mdtC, mexW,mexK), MLS (macA, macB), polymyxin (arnA, arnC), sulfonamide (sul1, sul2), and trimethoprim (dfrA3). Gene mac B (Macrolide resistance gene) was very predominant in nature. Several genera of antibiotic resistance bacteria have also been reported in activated sludge such as methicillin-resistant *Staphylococcus aureus*, vancomycin-resistant *Enterococcus* spp., and Gram-negative bacteria such as *Enterobacteria*, *Pseudomonas*, and *Acinetobacter* amongst others [55, 56]. Typically, members of Enterobacteriaceae reported resistance to 13 different antibiotics [57]. Apart from them opportunistic pathogenic bacteria such as *Enterococcus faecalis*, *Pseudomonas aeruginosa*, Enterobacteriaceae, *Staphylococcus aureus* and with antibiotic resistance gene was also reported [58].

5.7 Conclusion

Activated sludge is a process of biological treatment of wastewater which is widely practised all over the world. The entire process contains two separate phases which are the aeration phase and sludge settlement phase. The wastewater from the primary settlement tank is passed through the aerobic tank containing a wide range of microbial population. The aeration phase helps to maintain the microbial flocs and maximizes the contact between the surface of floc and the wastewater. Both culture dependent and culture independent assay methods were used to evaluate the microbial community of the activated sludge. The advent of the omics technology has helped us to have a thorough knowledge of the microbial community present in the activated sludge. These bacterial communities were repositories of many antibiotic resistance genes. Moreover, these microbial communities undergo several physiological functions, performs several types of biogeochemical cycles, and sequestration of nutrient from the sludge. A detailed understanding of the microbial community assembly will help us in understanding the ecological theories.

References

1 Scholz, M. (2016). Activated sludge processes. *Wetlands for Water Pollution Control* 91–105. doi: 10.1016/b978-0-444-63607-2.00015-0.
2 Huang, M.H., Li, Y.M., and Gu, G.W. (2010). a). chemical composition of organic matters in domestic wastewater. *Desalination* 262 (1–3): 36–42.
3 Park, H.D., Wells, G.F., Bae, H., Criddle, C.S., and Francis, C.A. (2006). Occurrence of ammonia-oxidizing archaea in wastewater treatment plant bioreactors. *Applied Environmental Science and Microbiology* 72: 5643–5647.
4 Lucker, S., Wagner, M., Maixner, F. et al. (2010). A *Nitrospira* metagenome illuminates the physiology and evolution of globally important nitrite-oxidizing bacteria. *Proceedings of NationalAcademy of Science, U S A* 107: 13479–13484.
5 Ye, L., Shao, M.F., Zhang, T., Tong, A.H., and Lok, S. (2011). Analysis of the bacterial community in a laboratory-scale nitrification reactor and a wastewater treatment plant by 454-pyrosequencing. *Water Research* 45: 4390–4398.

6 Geets, J., de Cooman, M., Wittebolle, L. et al. (2007). Realtime PCR assay for the simultaneous quantification of nitrifying and denitrifying bacteria in activated sludge. *Applied Microbiology and Biotechnology* 75: 211–221.

7 Jin, D.C., Wang, P., Bai, Z.H. et al. (2011). Analysis of bacterial community in bulking sludge using culture-dependent and -independent approaches. *Journal of Environmental Science* 23 (11): 1880–1887. doi: 10.1016/s1001-0742(10)60621-3.

8 Liu, B., Zhang, F., Feng, X. et al. (2006). *Thauera* and *Azoarcus* as functionally important genera in a denitrifying quinoline-removal bioreactor as revealed by microbial community structure comparison. *FEMS Microbiology Ecology* 55: 274–286.

9 Gao, P., Xu, W., Sontag, P. et al. (2016). Correlating microbial community compositions with environmental factors in activated sludge from four full-scale municipal wastewater treatment plants in Shanghai, China. *Applied Microbiology and Biotechnology* 100: 4663–4673.

10 Hu, M., Wang, X., Wen, X., and Xia, Y. (2012). Microbial community structures in different wastewater treatment plants as revealed by 454-pyrosequencing analysis. *Bioresource Technology* 117: 72–79.

11 Yu, K. and Zhang, T. (2012). Metagenomic and metatranscriptomic analysis of microbial community structure and gene expression of activated sludge. *PloS One* 7 (5): e38183. doi: 10.1371/journal.pone.0038183.

12 Gao, J., Luo, X., Wu, G., Li, T., and Peng, Y. (2014). Abundance and diversity based on amoA genes of ammonia-oxidizing archaea and bacteria in ten wastewater treatment systems. *Applied Microbiology and Biotechnology* 98: 3339–3354.

13 Pang, J., Matsuda, M., Kuroda, M. et al. (2016). Characterization of the genes involved in nitrogen cycling in wastewater treatment plants using DNA microarray and most probable number-PCR. *Frontiers in Environmental Science and Engineering* 10 (4): 07.

14 Siripong, S. and Rittmann, B.E. (2007). Diversity study of nitrifying bacteria in full-scale municipal wastewater treatment plants. *Water Research* 41: 1110–1120.

15 Zielinska, M., Rusanowska, P., Jarzabek, J., and Nielsen, J.L. (2016). Community dynamics of denitrifying bacteria in full-scale wastewater treatment plants. *Environmental Technology* 7: 1–10.

16 Mielczarek, A.T., Nguyen, H.T.T., Nielsen, J.L., and Nielsen, P.H. (2013). Population dynamics of bacteria involved in enhanced biological phosphorus removal in Danish wastewater treatment plants. *Water Research* 47: 1529–1544.

17 Mielczarek, A.T., Kragelund, C., Eriksen, P.S., and Nielsen, P.H. (2012). Population dynamics of filamentous bacteria in Danish wastewater treatment plants with nutrient removal. *Water Research* 46: 3781–3795.

18 Thomsen, T.R., Kong, Y., and Nielsen, P.H. (2007). Ecophysiology of abundant denitrifying bacteria in activated sludge. *FEMS Microbiology and Ecology* 60: 370–382.

19 Thomsen, T.R., Nielsen, J.L., Ramsing, N.B., and Nielsen, P.H. (2004). Micromanipulation and further identification of FISH labelled microcolonies of a dominant denitrifying bacterium in activated sludge. *Environmental Microbiology* 6: 470–479.

20 Nguyen, H.T.T., Nielsen, J.L., and Nielsen, P.H. (2012). '*Candidatus Halomonasphosphatis*', a novel polyphosphate-accumulating organism in full-scale enhanced biological phosphorus removal plants. *Environmental Microbiology* 14: 2826–2837.

21 Kristiansen, R., Nguyen, H.T.T., Saunders, A.M. et al. (2013). A metabolic model for members of the genus *Tetrasphaera* involved in enhanced biological phosphorus removal. *ISME Journal* 7: 543–554. doi: 10.1038/ismej.2012.136.

22 Nielsen, P.H., Nguyen, H.T.T., McIlroy, S., Mielczarek, A.T., and Seviour, R. (2013). *Identification of polyphosphate-accumulating and glycogen-accumulating organisms by FISH*. In: P.H. Nielsen.

23 Heylen, K., Vanparys, B., Wittebolle, L. et al. (2006). Cultivation of denitrifying bacteria: optimization of isolation conditions and diversity study. *Applied Environmental Microbiology* 72: 2637–2643.

24 Zou, H., Lu, X., and Saad, A. (2014). Characterization of microorganisms responsible for phosphorus removal linking operation performance with microbial community structure at low temperature. *Environmental Technology* 35: 1101–1109.

25 Zhang, H., Song, S., Jia, Y., Wu, D., and Lu, H. (2019). Stress-responses of activated sludge and anaerobic sulfate-reducing bacteria sludge under long-term ciprofloxacin exposure. *Water Research* 164: 114964.

26 Xu, Q., Liu, X., Wang, D. et al. (2018). Free ammonia-based pretreatment enhances phosphorus release and recovery from waste activated sludge. *Chemosphere* 213: 276–284.

27 Novo, A., André, S., Viana, P., Nunes, O.C., and Manaia, C.M. (2013). Antibiotic resistance, antimicrobial residues and bacterial community composition in urban wastewater. *Water Research* 47 (5): 1875–1887.

28 Ye, L., Mei, R., Liu, W.T. et al. (2020). Machine learning-aided analyses of thousands of draft genomes reveal specific features of activated sludge processes. *Microbiome* 8: 16. doi: 10.1186/s40168-020-0794-3.

29 Yang, Y., Wang, L., Xiang, F., Zhao, L., and Qiao, Z. (2020). Activated sludge microbial community and treatment performance of wastewater treatment plants in industrial and municipal zones. *International Journal of Environmental Research and Public Health* 17 (2): 436. doi: 10.3390/ijerph17020436.

30 Ai, C., Yan, Z., Zhou, H. et al. (2019). Metagenomic insights into the effects of seasonal temperature variation on the activities of activated sludge. *Microorganisms* 7 (12): 713.

31 Zhao, R., Feng, J., Yin, X. et al. (2018). Antibiotic resistome in landfill leachate from different cities of China deciphered by metagenomic analysis. *Water Research* 134: 126–139.

32 Guo, J., Ni, B.-J., Han, X. et al. (2017). Data on metagenomic profiles of activated sludge from a full-scale wastewater treatment plant. *Data in Brief* 15: 833–839. doi: 10.1016/j.dib.2017.10.048.

33 Xia, Y., Hu, M., Wen, X. et al. (2016). Diversity and interactions of microbial functional genes under differing environmental conditions: insights from a membrane bioreactor and an oxidation ditch. *Science Report* 6: 18509. doi: 10.1038/srep18509.

34 Shchegolkova, N.M., Krasnov, G.S., Belova, A.A. et al. (2016). Microbial community structure of activated sludge in treatment plants with different wastewater compositions. *Frontiers in Microbiology* 7: 90. doi: 10.3389/fmicb.2016.00090.

35 Yadav, T.C., Khardenavis, A.A., and Kapley, A. (2014). Shifts in microbial community in response to dissolved oxygen levels in activated sludge. *Bioresource Technology* 165: 257–264. doi: 10.1016/j.biortech.2014.03.00.

36 Wang, H., Qun, S., Jing, W. et al. (2018). Simultaneous nitrification, denitrification and phosphorus removal in an aerobic granular sludge sequencing batch reactor with high dissolved oxygen: effects of carbon to nitrogen ratios. *Science of Total Environmental* 642: 1145–1152.

37 Heylen, K., Lebbe, L., and De Vos, P. (2008). *Acidovoraxcaeni* sp. nov., a denitrifying species with genetically diverse isolates from activated sludge. *International Journal Systematics and Evolutionary Microbiology* 58: 73–77.

38 Chen, D.Y., Wang, Z.M., Zhang, M.L., Wang, X.H., and Lu, S. (2019). Effect of increasing salinity and low C/N ratio on the performance and microbial community of a sequencing batch reactor. *Environmental Technology* 31: 1–12.

39 Huang, Z., Gedalanga, P.B., Asvapathanagul, P., and Olson, B.H. (2010b). Influence of physicochemical and operational parameters on *Nitrobacter* and *Nitrospira* communities in an aerobic activated sludge bioreactor. *Water Research* 44: 4351–4358.

40 Meng, F., Gao, G., Yang, T.-T. et al. (2015). Effects of fluoroquinolone antibiotics on reactor performance and microbial community structure of a membrane bioreactor. *Chemical Engineering Journal* 280: 448–458.

41 Terashima, M., Yama, A., Sato, M. et al. (2016). Culture-dependent and independent identification of polyphosphate-accumulating *Dechloromonas* spp. predominating in a full-scale oxidation ditch wastewater treatment plant. *Microbes and Environment* 31: 449–455.

42 Konneke, M., Bernhard, A.E., de la Torre, J.R. et al. (2005). Isolation of an autotrophic ammonia-oxidizing marine archaeon. *Nature* 437: 543–546.

43 Jiang, R., Wang, J.G., Zhu, T. et al. (2020). Use of newly designed primers for quantification of complete ammonia-oxidizing (comammox) bacterial clades and strict nitrite oxidizers in the genus *Nitrospira*. *Applied and Environmental Microbiology* 86 (20): e01775–20. doi: 10.1128/AEM.01775-20.

44 Karkman, A., Do, T.T., Walsh, F., and Virta, M.P. (2017). Antibiotic-resistance genes in wwastewater. *Trends of Microbiology* 26 (3): 220–228.

45 Martinez, J.L. (2009). Environmental pollution by antibiotics and by antibiotic resistance determinants. *Environmental Pollution* 157 (11): 2893–2902.

46 Nnadozie, C., Kumari, S., and Bux, F. (2017). Status of pathogens, antibiotic resistance genes and antibiotic residues in wastewater treatment systems. *Review in Environmental Science and Biology* 16 (3): 491–515.

47 Mao, D., Yu, S., Rysz, M. et al. (2015). Prevalence and proliferation of antibiotic resistance genes in two municipal wastewater treatment plants. *Water Research* 85: 458–466.

48 Liu, Z., Klümper, U., Liu, Y. et al. (2019). Metagenomic and metatranscriptomic analyses reveal activity and hosts of antibiotic resistance genes in activated sludge. *Environment International* 129: 208–220. doi: 10.1016/j.envint.2019.05.036.

49 Yang, Y., Li, B., Zou, S., Fang, H.H., and Zhang, T (2014). Fate of antibiotic resistance genes in sewage treatment plant revealed by metagenomic approach. *Water Research* 62: 97–106.

50 Bengtsson-Palme, J., Hammarén, R., Pal, C. et al. (2016). Elucidating selection processes for antibiotic resistance in sewage treatment plants using metagenomics. *Science Total Environmental* 572: 697–712.

51 Karkman, A., Johnson, T.A., Lyra, C. et al. (2016). High-throughput quantification of antibiotic resistance genes from anurban wastewater treatment plant. *FEMS Microbiology Ecology* 92 (3).

52 Pal, C., Bengtsson-Palme, J., Kristiansson, E., and Larsson, D.J. (2016). The structure and diversity of human, animal and environmental resistomes. *Microbiome* 4 (1): 54.

53 Van Goethem, M.W., Pierneef, R., Bezuidt, O.K. et al. (2018). A reservoir of 'historical' antibiotic resistance genes in remote pristine Antarctic soils. *Microbiome* 6 (1): 40.

54 Yang, Y., Li, B., Ju, F., and Zhang, T. (2013). Exploring variation of antibiotic resistance genes in activated sludge over a four-year period through a metagenomic approach. *Environmental Science and Technology* 47 (18): 10197–10205.

55 Bouki, C., Venieri, D., and Diamadopoulos, E. (2013). Detection and fate of antibiotic resistant bacteria in wastewater treatment plants: a review. *Ecotoxicology Environmental Safety* 91: 1–9. doi: 10.1016/j.ecoenv.2013.01.016.

56 Figueira, V., Vaz-Moreira, I., Silva, M., and Manaia, C.M. (2011). Diversity and antibiotic resistance of aeromonas spp. in drinking and wwastewater treatment plants. *Water Research* 45: 5599–5611. doi: 10.1016/j.watres.2011.08.021.

57 Amador, P.P., Fernandes, R.M., Prudencio, M.C., Barreto, M.P., and Duarte, I.M.(2015). Antibiotic resistance in wastewater: occurrence and fate of Enterobacteriaceae producers of class A and class C β-lactamases. *Journal of Environmental Science Health A Tox. Hazard. Subst. Environ. Eng.* 50 (1): 26–39. doi: 10.1080/10934529.2015.964602.

58 Alexander, J., Bollmann, A., Seitz, W., and Schwartz, T. (2015). Microbiological characterization of aquaticmicrobiomes targeting taxonomical marker genes and antibiotic resistance genes of opportunistic bacteria. *Science of Total Environment* 512–513: 316–325. doi: 10.1016/j.scitotenv.2015.01.046.

6

Pre-treatment of Industrial Wastewater

Vaishnavi Sharma[1] and Satarupa Dey[2,]*

[1] *National Institute of Technology Warangal, Telangana, India*
[2] *Department of Botany, Shyampur Siddheswari Mahavidyalaya, Howrah, West Bengal, India*
* *Corresponding author*

6.1 Introduction

6.1.1 Background

Manoeuvring industrial waste is a crucial challenge for all operational and non-functional industries around the world. The industrial revolution carelessly generated varieties of waste as its by-product, which is now difficult to treat and dispose of. Industrial waste can be largely characterized into four categories (Figure 6.1). Water is a scarce resource and its use is being checked to maintain its availability for everyone. Almost every industry requires water for various purposes and thus their production of wastewater is inevitable. Industries tend to produce different types of wastewaters such as [1]; (a) sewage water, which comes out from industries; (b) storm water, which collects in the vicinity of the industry. Its impurity levels are defined by the amount of pollution in that particular catchment area; (c) technological water, which emerges as the end product from different processes; (d) cooling water which is often circulated within the system; (e) water emerging from energy units, contaminated with high levels of dissolved minerals. Due to increasing societal pressure and environmental concern, industries are being compelled to treat the wastewater they produce before its disposal. Involving the circular economy and sustainability concepts, wastewater is seen as a potential resource as its recycling and treatment can generate various economic and financial benefits. The main aim of this chapter is to analyse the pre-treatment procedures currently being followed in various industries. The possible advancements are also suggested in a later study. The main objectives are as

Application of Sewage Sludge in Industrial Wastewater Treatment, First Edition. Edited by Maulin P. Shah.

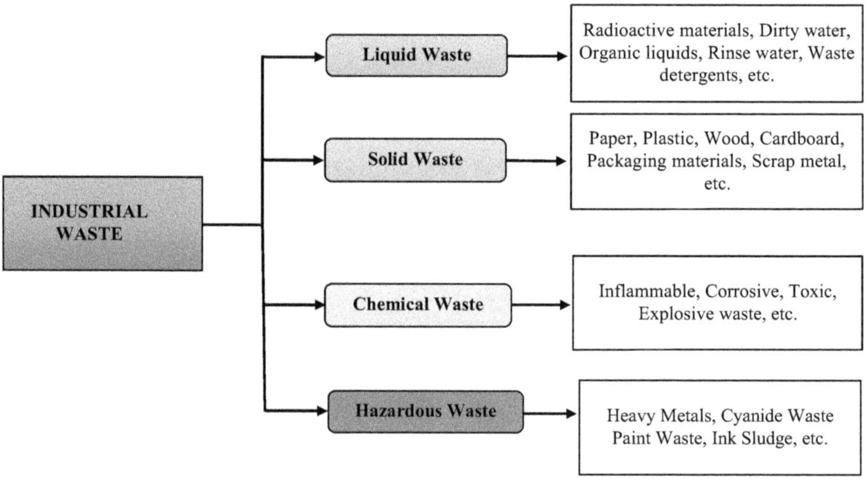

Figure 6.1 Types of industrial waste.

follows: (a) to understand the mechanism of processes involved in pre-treatment of industrial wastewater; (b) to use this knowledge to suggest more ways of efficient treatment; (c) to explore the application of these new technologies to different processes in diverse industries. The study involved the investigation of works involving the pre-treatment of industrial wastewater being applied in different industries presently, e.g., pre-treatment of wastewater from pulp mills, the petrochemical industry, etc. In addition, all the factors influencing the selection of pre-treatment processes in different industries were studied. Later a comprehensive analysis was carried out to compare the different results from various pre-treatments provided to diverse wastewaters and a knowledge of that was used to suggest some pre-treatments to new under-explored industries producing different wastes. The following limitations of the current study can be listed as follows. (a) The study was intended as a preliminary investigation of multiple factors affecting various pre-treatments in order to obtain a more stable substrate. Therefore, this was done by studying the works done to date. The results of the analysis are aimed to be interpreted, at this stage, qualitatively, not quantitatively. (b) In order to find out the actual mechanism of processes involved, the study relies on the already available results from various research. In this chapter, a comparative study was carried out on various pre-treatment practices carried out by different researchers on waste samples from different industries. This critical review explains the scope of application of these treatments to wastewater from distinct industries. Even though the concept was introduced many years ago, its applications are still unexplored for contrasting sectors producing wastewater on a large scale.

6.2 Importance of Key Industrial Sectors

Industrialization started in the early 1980s and over the course of time has changed the face of the world. The benefits that it has brought to society are countless yet its menacing effects cannot be tolerated. The most important industries that work as the backbone in running the world are discussed below. Newspapers and books have been in existence for many years. They are highly important for readers and for manufacturers. The contribution of the paper and pulp industry ranks highly in the world, has existed in various forms and can be clearly seen frequently in our day to day life. The coal industry has always been a major power source which keeps the world moving. The petrochemical industry also plays a vital role; the plastics produced by it are used by all industries including automobile, agriculture, etc., for various purposes like packaging, moulding, insulation etc. It also makes major contributions to the nation's economy [2]. The textile industry is involved in the creation of yarn from various sources like sheep and cotton which are later used as a raw material for the creation of clothes fulfilling one of the basic needs of people. Good health care facilities are a necessity for every section of a society. The development of designs and production of drugs for medication purposes, which can be life savers are part of the pharmaceutical industry. Olive oil is considered to be an excellent edible oil. It has innumerable benefits: it is rich in antioxidants, it helps control cholesterol levels, aids in preventing many heart diseases and strokes, it also has anti-cancer properties, controls diabetes and Alzheimer's disease, etc. Production of olive oil is carried out by oil industries from many small and large countries. It also supports the GDP of these countries. Food is a requirement for all living beings. The food we get on our tables start its journey on farms and in the hands of farmers. Hence, the fertilizer industry is at the peak which fulfils the requirements of all kinds of modern agricultural practices. From the air we breathe, to the food we eat, to the things we use, etc., all come from one or other of these industries. They affect everything around us. And hence their proper working is essential for our well-being.

6.3 Need for Pre-treatment of Industrial Effluents

Water, being a scarce resource, needs to be used economically and sustainably. All the major industries require water for its operations and thus production of wastewater is essential for them. Water is used by paper industries (approximately 30 m^3 per ton), sugar industries (approximately 10 m^3 per ton), and dying industries (approximately 100 m^3 per ton) [3]. The uncontrolled industrialization has imposed an urgent need to recover usable water from already used resources. The need in the dying industry to prevent environmental impact and to match the effluent discharge quality standards paved the way for implementing new and advanced treatment technologies. The paper and pulp industry is found to use the highest amount of cellulosic resources, and once the resources are used the challenge to bring them back into the value chain is critical. Similarly, coal being a non renewable energy, once used it takes years to replenish, but coal is still utilized as a major power source for many industries. The industrial waste from the coal industry can be used for zeolite synthesis, CO_2 capture and the manufacture of adsorbents only when given a specific pre-treatment [4]. Despite all the benefits we receive from different industries, there is an immediate need to address the pollution being caused by them. Hence, we require improved hybrid methods to ensure good quality standards of discharge from industries. Waste generation can only be reduced, not avoided, by the use of new techniques. As the demand for products is increasing day by day to meet the requirements of the people, production cannot be stopped. The wastewater coming from the industries have varied contents. The characteristic of wastewater produced depend on the type of industry and the processes involved. The pollutants mainly present are heavy metals (coal industry), hazardous chemicals (textile industry), polyaromatic hydrocarbons and heterocyclic compounds (petrochemical industry), etc. All these pollutants are extremely harmful to human health and the environment. The harm that they cause can be listed from mild to severe impacts on humans like; nausea, headache, skin irritation, and serious issues like cancer, birth defects, neurological impairments, reproductive anomalies [4]. The level of toxicity of the contents in the discharged wastewaters from several industries has reached 10^{12} revertant (in *Salmonella*) per litre which is alarmingly high. This kind of genotoxicity harms all kinds of living organisms and also associated with mutagenic hazards [4]. Wastewater, when it is disposed of to water bodies from commercial and industrial sources gives a mixed polluted substrate which creates havoc for aquatic life and people using that water directly or indirectly. Moreover, the solid waste produced from these industries, when disposed into landfills, contributes to leachate formation which again pollutes the groundwater causing permanent damage. On the other hand, if these compounds are treated by different processes, their recovery becomes easier and when they are disposed of create fewer issues. This involves the need to apply pre-treatment processes to improve the final results.

6.3.1 Pre-treatment Processes

Different industries produce a diversified range of effluents. Due to the presence of a huge amount of pollutants in the discharged water, it is a very difficult task to even identify all the components present in it [3]. The establishment of the pre-treatment processes is evidence that portrays the deficiency of many existing wastewater treatment processes. The effluents discharged are highly complex in nature and if just given a single treatment the results are not good enough, economic analysis shows that they are expensive and the end products require further treatment [4]. Pre-treatment processes are used worldwide to ensure that the discharge standards of the disposed water are checked and the processes involved in the treatment become more environment-friendly [4]. The pre-treatment processes should be selected in such a manner that when the secondary treatment is carried out on the substrate the end product meets the discharge standards without requiring any further care. Environmental concerns are the major factors which impelled the study of advanced pre-treatment technologies as of high priority globally [4]. A successful pre-treatment technology results in the reduction of all harmful contents, physio-chemical parameters before the wastewater is subjected to secondary or advanced treatment.

6.3.2 Sustainability Approach

Society is slowly becoming aware of the circular economy and sustainability concepts. Awareness has been boosted because of the sudden environmental changes which have caused catastrophes around the world. The application of the 10R's (Reuse, Reduce, Recycle, Recover, Refill, Refuse, Repair, Repurpose, Rethink, Refurbish) has huge scope in handling industrial waste as it has great potential in generating economy in an eco-friendly way. Although the wastewater treatment processes use resources like electricity, etc., for their working, they do have a potential to pay back to society in other forms [5]. The zero-liquid discharge concept is another game changer concept in reducing waste of all forms from industries. When this concept is applied to brine solution emerging from the desalination of substrate, the liquid is treated and re-useable water and salts are effectively acquired [6].

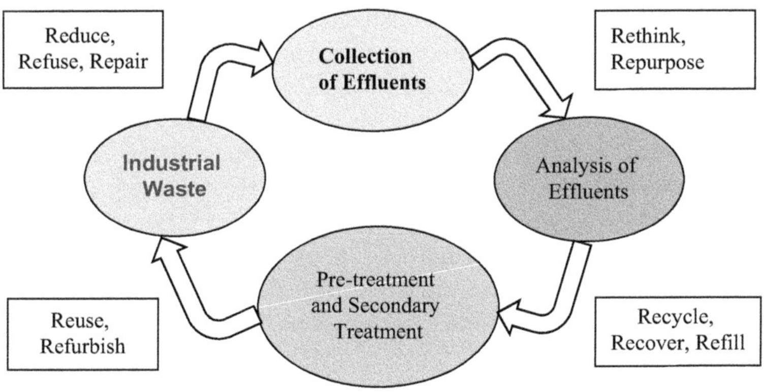

Figure 6.2 Schematic representation for sustainability approach in handling industrial waste.

6.4 Classification of Pre-treatment Technologies

6.4.1 Classical Technologies

The pressure on industries to treat the wastewater being produced by them has been growing over the years. There are several studies available from different researchers from around the world suggesting ways to treat the wastewater produced. The importance of pre-treatment technologies is presented by the extent to which it handles and treats waste before it can be supplied for the main secondary treatment. Physical pre-treatments have found a significant place in laboratory experiments and in the industries too. Thermal pre-treatments are largely used to release the intracellular compounds by devastating the cell walls and membranes of the microorganisms [4]. One of the most common universally used physical treatment is ultrasonication. In this process, the molecular cell-wall breaks and forms holes exposing the intracellular materials to fast dissociation. The hydroxyl ion formation also helps in degradation [4]. This system is preferred as it has economic advantages over the ozonation process. Biological approaches to treat wastewater from industries are also considered important keeping in mind the enzymatic hydrolysis concept [4]. Chemical pre-treatments like Advanced oxidation processes (AOP) have also been largely used to treat wastewater from textile industries to handle the organic content present in it. In AOP A strong oxidizing radical, a hydroxyl ion, is present which breaks the organic compounds into carbon dioxide, water and inorganic ions which can be handled separately later [4]. Ozonation is a common but a costly technique. Its effectiveness depends on the extent to which it can create hydroxyl ions and alter the unsaturated bonds present in the substrate. One of the most frequently used and studied technique is wet air oxidation (WAO). In this technique, under high temperature and pressure the organic compounds are broken down to CO_2, H_2O and inorganic compounds which are separated and handled individually [4]. Radiation pre-treatment is again a highly used technology found in several industries. This is preferred as its use eliminates the requirement of any further treatment. In this process, non-biodegradable compounds are completely degraded and can be disposed of easily later [4]. Coagulation and flocculation are conventional units used for pre-treatment in industries having oil and grease mixed in their substrates. They are used as pre-treatments because they are highly effective in reducing fouling of membranes used in the secondary treatment units [4]. Zero valant ion pre-treatment is a strong oxidizing reagent which ensures the movement of non-toxic compounds to secondary treatment units for further biological measures [4].

6.4.2 Hybrid Methods

The evolution of pre-treatment processes took place in order to gain an econor-cal edge over the treatment procedures applied. The need to convert from the lo established pre-treatment procedures to more advanced, modern procedures a came because of increasing environmental pollution. Effluent discharges are increasing gradually because of the uncontrolled production on large scales. This imposed the need for more research into techniques to handle this huge volume of waste. Thus, the formation of hybrid or mixed pre-treatment techniques came into existence. These technologies not only prevent pollution but also could generate economy through recovered resources from the waste. The adsorption technique is used with coagulation and flocculation to ensure less membrane fouling. The nanoparticles are great composites to be used in the coagulation and flocculation units as they have the ability to create large and dense settleable particles which are easily removed with the sludge. The nanoparticles are small in size but have large surface area and also readily react with the effluents which make them favourable for use [4]. Thermal-hydrolysis when carried out with alkaline treatment requires low energy input and overall is cost efficient because the cost of reagents, etc., can be easily avoided. Thermal-hydrolysis can be combined with AOP's in a similar manner [7]. Microwave irradiation accompanied by alkali treatment as its secondary treatment is also a favourable technique according to various studies. Ultrasonic techniques can be used in combination with acid/alkali reactions, microwave, ozone etc. [7]. For better bio-methane production ultrasonication and microwave treatment when combined together give better results according to some researchers [4]. Fenton oxidation can be carried out by ultrasonic or bio-leaching methods to obtain better results. When Fenton is used with ultrasonic treatment, the nutrients like nitrogen, potassium and phosphorous are made available in the substrate which are beneficial for further biological treatment like composting [7]. There are studies portraying the effectivity of the use of such combined hybrid processes on a laboratory scale. The economic analysis is yet to be carried out for such combinations, but with the availability of present data, it can be easily done on demand by the isadifferent owners of the industries.

6.4.3 Categorization of Techniques

The wastewater coming out of the industries can be treated by various methods. The ultimate aim of these treatments is to treat the substrate in such a way that it renders zero harm to the environment. Physical treatments include ultrasonic pre-treatment, thermal pre-treatment, microwave pre-treatment, lysis-thickening centrifuge, stirred ball mill, high pressure homogenization and focused pulsed pre-treatment. The technologies are broadly classified as shown below.

6.4.3.1 Thermal Pre-treatments

Thermal treatments are quite expensive at industrial scale so they are just applied at pilot scale currently. The treatment involves breaking of cell walls of organic components thus initiating fast degradation. Thermal treatment can effectively reduce the sludge quantity produced and also hasten the process to a good extent [8].

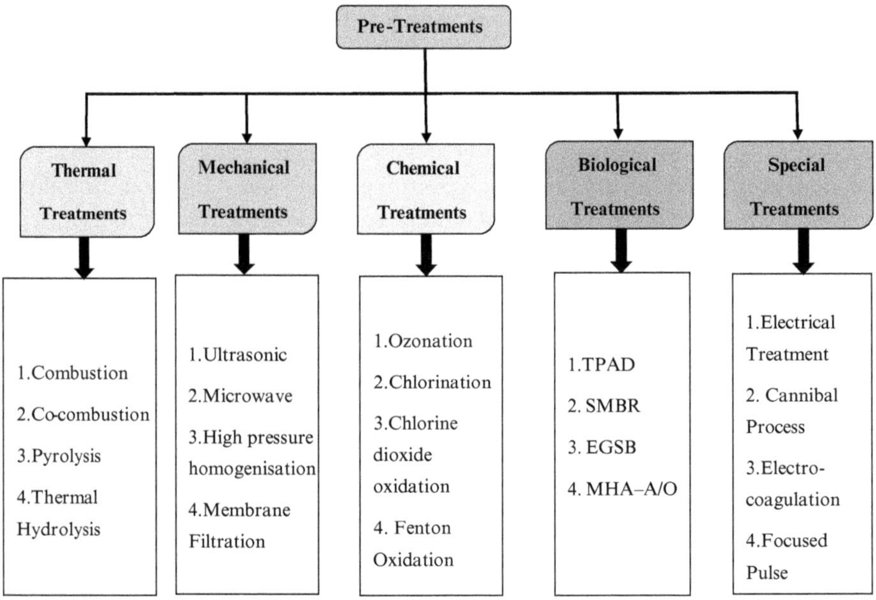

Figure 6.3 Classification of pre-treatment technologies.

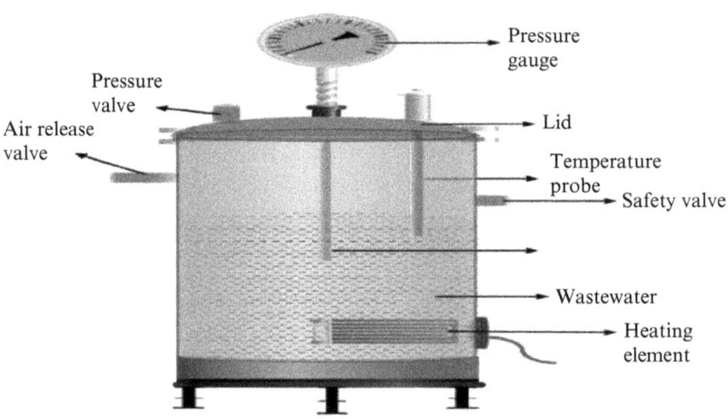

Figure 6.4 Thermal pre-treatment setup [10].

Thermal treatment is usually used for wastewater from the olive oil industry. Combustion and pyrolysis is a common practice; although used for energy recovery the emissions from these processes are toxic gases which harm the environment. So, to in order to control that more accessories are required accompanying the remaining processes [9]. Thermal hydrolysis is preferred as it affects biological processing of secondary treatments in such a way that it improves the bio-gas production potential, lowers the hydraulic retention time of the substrate and also effectively removes pathogens. This treatment is found to lower the viscosity and improve the dewaterability of the substrate, eventually causing the reduction in the volume of the end product. The drawback of this treatment is that the non-biodegradable content of the substrate remains unchanged [7].

6.4.3.2 Mechanical Pre-treatments

Mechanical pre-treatments require instruments to carry out their operations. They are pre-requisites for any treatment plant as they can reduce the maintenance cost of the plant [1]. Some traditionally used pre-treatment techniques are as follows. (a) Bar screens: the main purpose of bar screens is to remove floating objects which may clog the treatment units further. The trapped material is separated out manually or mechanically [1]. (b) Sand traps: gravel or sand particles of size 0.2 mm are removed in this unit. The basic principle behind this is to reduce the flow rate while the wastewater is sent to a sedimentation tank [1]. (c) Imhoff tank: the main aim is to remove fine sludge particles by the use of a deeper sedimentation tank. The maintenance for these tanks is minimal, they require cleaning twice a year, are very reliable and useful [1]. With the passage of time, these traditional techniques were not able to meet the demand of the market on cost effectiveness and the amount of time consumed. This led to the formulation of new and advanced techniques which were faster and also improved the efficiency of the overall system [1]. Ultrasonic process works on the principle of cavitation and produces hydroxyl ions as its end product. It is a two-step process, the first one being floc-losing and next cell-breaking [7]. Hydroxyl ions help in deconstructing the wastewater sludge. During cavitation, millions of molecules in the substrate collapse and produce high temperatures and pressures, also high shear forces which together destroy the substrate structure [8]. This treatment has been applied largely in industries worldwide after being successfully tested in laboratories. The limitation of this process is that it can just hasten the degradation process but cannot reduce the sludge produced as the end product [8]. Microwave pre-treatments are yet to be applied in industry. It has, however, been successfully tested in laboratories. This process can effectively reduce the volume of sludge produced and can also do it in less time [8]. The high pressure homogenizer is an piece of equipment used to tackle industrial waste sludge. In this technology, the speed of the substrate is increased so that collision of molecules takes place accompanied by cavitation which dissociates the sludge as a result [8]. Microwave irradiation is a more often used method as compared to traditional heating methods. A number of advantages can be listed as; rapid and selective heating; instant control of processes; compactness; and cost effectiveness, etc. Microwave irradiation can be carried out in two ways, i.e., low temperature and high pressure combination and high temperature and low pressure combination [7]. Considerable care is required for this treatment because, if the temperature and pressure are not maintained it results in the solubilization of the substrate, which will eventually increase the operational and maintenance costs making the process unfit for use. The main reason it is used is because the additional cost is balanced by the bio-gas generation quantity, a fuel in demand for the future. Reverse osmosis (RO) is carried out as a secondary treatment for various industries. The drawbacks of just using single step RO or nano filtration (NF) is that the system tends to foul due to clogging of different suspended solids, organic materials and bacteria, scaling is also caused by the presence of ions in wastewater [11]. The major challenge for RO treatment is fouling of membranes. Fouling causes issues like irreversible membrane damage, reduced flux rates and increased operating costs, to avoid this to a greater extent pre-treatments are carried out [11]. Membrane Infiltration is a common and most attractive pre-treatment technique. Due to its simplicity in operation it is applied largely to improve treatment efficiency [12]. Membrane filtration is carried out to mostly tackle the organic waste components found in the industrial wastewater. If a proper combination of pre-treatment techniques are chosen, then the permeate formed may be free of suspended solids (SS) [12]. With the best combination applied the cleaning and maintenance requirement of the plant decreases which will positively change the productivity and hence will make the system more economically viable. The productivity can be doubled with a proper pre-treatment [12]. Membrane filtration can be carried out by some common methods; (i) membrane ultrafiltration, (ii) dual media filtration and granular activated carbon (GAC) adsorption, (iii) dual media filtration with dosage of organic flocculent and GAC adsorption. The flocculent quantity should be properly measured in order to avoid it clogging in the RO unit [12]. Microfiltration and ultrafiltration are common practices in industry. Ultrafiltration is one of the most reliable techniques as it has advantages like;

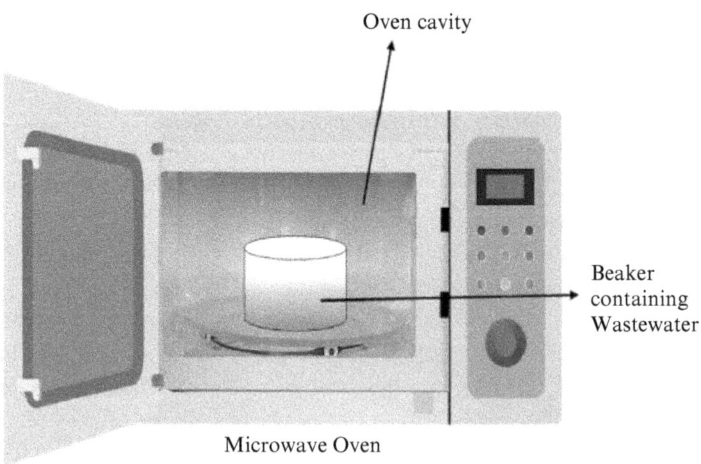

Figure 6.5 Microwave pre-treatment setup [10].

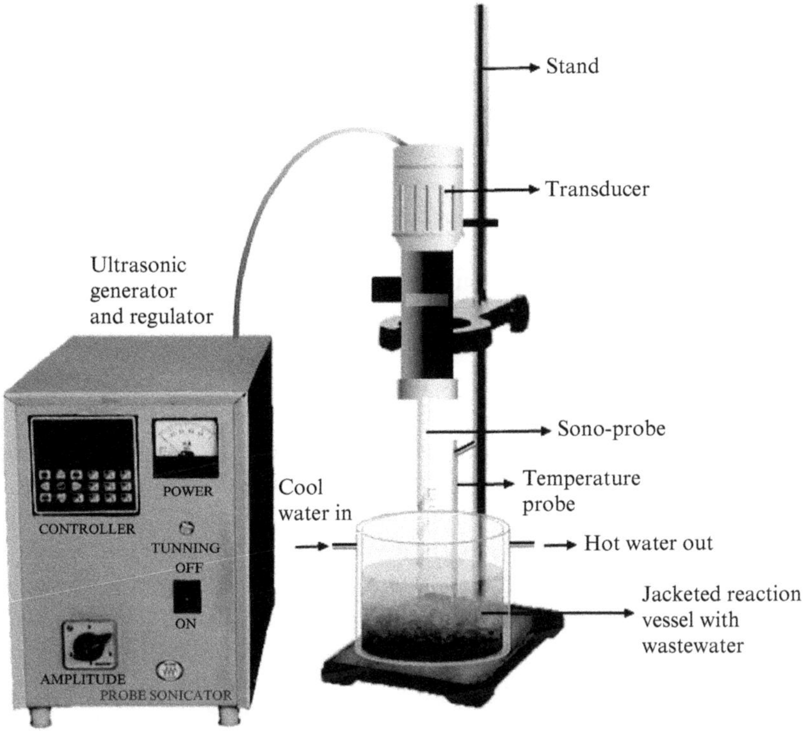

Figure 6.6 Ultrasonic pre-treatment setup [10].

outstanding water quality, efficient energy consumption, small footprint and easy operation [13]. Ultrafiltration coupled with biological treatments effectively avoids fouling of membranes thus improving the process efficiency [14]. Continuous microfiltration (CMF) is also a promising pre-treatment technology applied before RO treatment. It has wide application for industrial and municipal wastewater. The major issue during its operation is maintaining the substrate's integrity, i.e., ensuring that there is no leakage from the removed matter in the end-product [11].

6.4.3.3 Chemical Pre-treatments

Chemical treatments are largely applied by industries like paper and pulp, textiles and tanneries. This is due to the fact that they require numerous chemicals for their production. The usual chemical treatments that are carried out by them are Fenton, coagulation, adsorption and membrane technology [13]. Treatments involving oxidation and free nitrous acid (FNA) treatment come under this category. Oxidation is primarily carried out by ozonation which is a very costly process. A few other treatments commonly utilized are chlorination, chlorine dioxide oxidation and Fenton oxidation [8]. Ozonation has been very popular since its introduction in the mid 1990s. It is highly advantageous as it can effectively disintegrate sludge and cause cell destruction of various compounds. It also affects the solubilization and mineralization of various compounds and particulates residing in the substrate [8]. Fenton oxidation belongs to the advanced oxidation processes (AOP) [13]. In Fenton oxidation, the destruction of organic components, i.e., macromolecules, are carried out by hydroxyl ion radicals which is the end product of the reaction between H_2O_2 and Fe^{2+} ions [8,13]. The main reason behind its large application is due to the involvement of iron in its process. The procedure can be understood in four parts; (1) the adjustment of pH to low acidic condition; (2) oxidation; (3) neutralization; and (4) by-products separated by coagulation [7]. Iron on the other hand, being a non-toxic and cheap element makes the treatment more favoured [7]. Fenton treatment can easily tackle complex molecules having high molecular weights and hence is largely preferred. It is found that the lowest flux reduction is found by this method [13]. Fenton, when given as a pre-treatment before radiolysis, is found to remove silica from the waste stream [13]. Membrane fouling is mostly attributed to bio-polymer fraction in the wastewater, which is effectively removed by Fenton reagents, ensuring smooth treatment with reduced cost and time of operation. Ferrous nitrous acid (HNO_2), has also been tried at pilot scale initially as it has proved to reduce the sludge production in laboratory tests. When nitrification of anaerobic sludge digestion liquor is carried out, FNA can be recovered from the treatment unit as its by-product and therefore, making it a renewable chemical [8]. AD when given a pre-treatment of FNA improves bio-methane formation [7]. The combination of heating or thermal treatment accompanied with FNA are found to produce good results in the laboratory. Sludge reduction is the main aim of these oxidation processes. Ozonation can reduce it by 10% in volume or can even reduce it to zero. Chlorine, chlorine dioxide and Fenton reagents can also effectively oxidize the components resulting in sludge reduction. Although they are still applied and tested just at laboratory scale and are yet to be implemented on a large industrial scale [8]. Catalytic wet oxidation is another process in which organics are converted into CO_2, H_2O in aerobic conditions. This reaction is carried out in liquid phase and the catalyst is used here to fasten the overall process. If this process is aided by carbon supplements, then the rate of biological nitrification is increased resulting in the removal of organics from the waste stream [7]. Sludge degradability can also be improved by alkaline treatment. This process enables breaking of carbohydrates, lipids and proteins into smaller compounds which are readily soluble and can be further treated. Although it still waits to be applied at industrial scale [8]. Coagulation is a traditional technique applied to remove organic components from the wastewater. Usually coagulation is carried out be[illegible]

Reverse osmosis (RO) or Ultrafiltration (UF) to remove suspended and colloidal particles from the substrate [13]. The feed after coagulation when moved for secondary treatment is found to show less fouling as it removes algal organic matter to a good extent. Photo-catalytic pre-treatment has wide usage. It not only protects the environment from any harm but also fulfils the energy requirements at low production cost. Contaminant removal, pathogen excretion and deletion of toxic components of the substrate is done throughout the process, resulting in reduced solubilization and higher energy production [7].

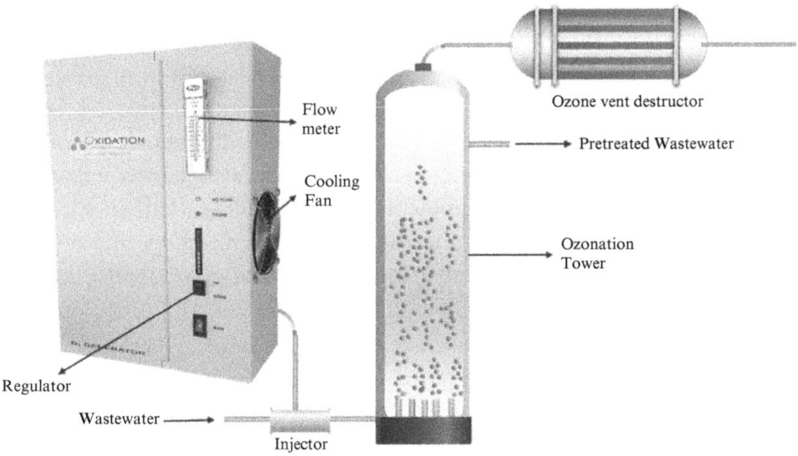

Figure 6.7 Ozone pre-treatment setup [10].

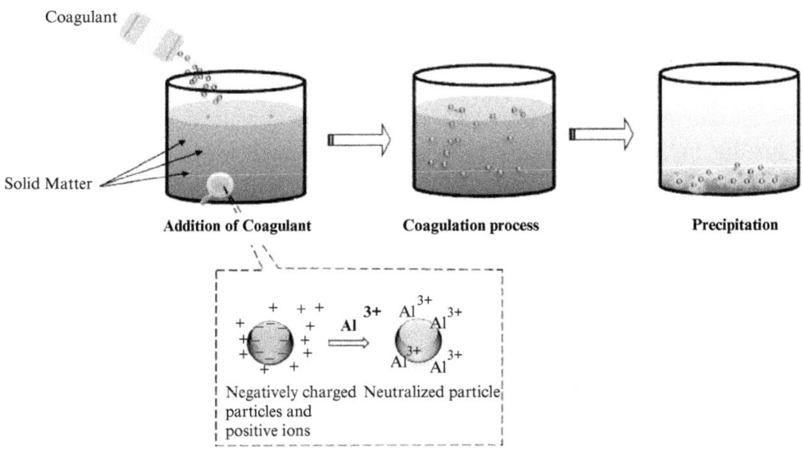

Figure 6.8 Coagulation-flocculation pre-treatment process [10].

6.4.3.4 Biological Pre-treatments

Temperature-phased anaerobic digestion (TPAD) is usually used in industry. The pre-treatment is given to the substrate in order to improve the efficiency of the AD process. Conventional the AD process is divided into the four stages; hydrolysis, acidogenesis, acetogenesis, methanogenesis. The first stage is considered to be the preliminary stage and requires less retention time and the following stages require more time to function. With the aid of TPAD, the first three stages are separated from the last one, facilitating better maintenance and improved process speed [8]. The use of anaerobic digestion as a pre-treatment is also carried out in treating wastewater having constructed wetlands as the secondary treatment. This combination has found a lot of application in the field as it has various advantages; lower maintenance costs due to less clogging of reactor beds, the entire setup can be used for years without any issue, it is an inexpensive method, etc. [15]. Other biological processes can be listed accordingly as; submerged membrane bioreactor (SMBR), micro-aerobic hydrolysis-anoxic/oxic processes, anaerobic expanded granular sludge bed (EGSB), micro-aerobic hydrolysis–acidification–anoxic–oxic processes (MHA–A/O) [2]. All these processes are able to reject specific organic toxic pollutants at high temperature, ensuring fast degradation of organic matter in the substrate [2].

6.4.3.5 Special Pre-treatments

Electrical treatment directly affects the phospholipids in the cell membranes of organic matter, and thus compels them to disintegrate to form a stable substrate. With the start of the treatment, there is a formation of cell pores which are accompanied by cell rupture. Due to this, the internal intracellular materials are exposed to electricity and further break down into smaller, more stable forms. In this manner subsequent degradation is carried out [8]. Electro-coagulation has been extensively studied researchers to date. It is similar to chemical coagulation [16]. In this process, coagulation is carried out in the presence of iron or aluminium electrodes. Coagulant is produced from one of the electrodes used in the cell. The process results in reduced sludge formation and lowers dewaterability [7,9]. Focused pulse treatment is a pioneering technology in which rapid pulse at high voltage is used in order to carry out the breakage of cell membranes of heavy molecules and organic pollutants. In the process, the substrate is introduced to high energy-pulsed electric field (PEF) wherein the entire bio-degradation takes place [7]. Freezing and thawing is another pre-treatment technique which carries out irreversible changes in the structure of the sludge. Although it lowers dewaterability to a good extent it also consumes a lot of energy making it unaffordable [7]. A cannibal process was introduced in 2008, it is a patented process and involves the particular combination of physical and biological treatments [8]. This process effectively reduces the sludge from wastewater.

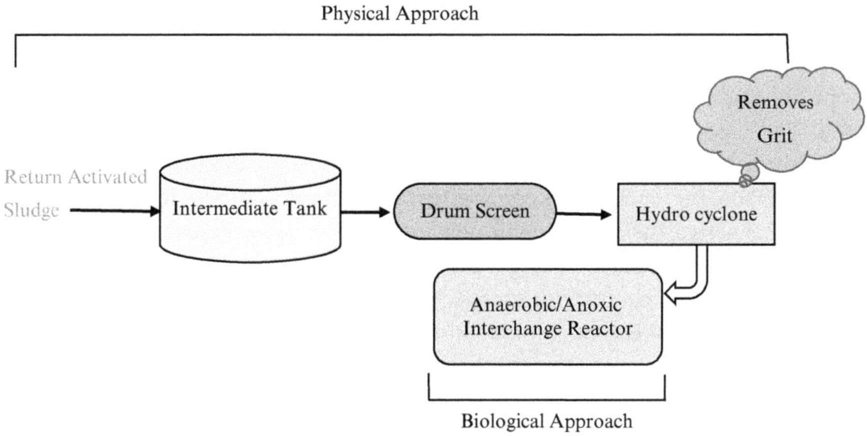

Figure 6.9 Cannibal process approach.

6.5 Characteristics of Different Pre-treatment Technologies

6.5.1 Biodegradability

The industrial wastewater is mostly treated through biological treatments but for this to happen the wastewater must possess some typical physiochemical characteristics and biological contaminants beforehand. If the bio-degradability index is 0.4, then these processes can be readily applied; photo-Fenton oxidation, chemical oxidation, ZVI pre-treatments, catalytic ozonation etc. [4].Electro-chemical treatments can also be used to degrade certain compounds in wastewater efficiently. Although, there are ample studies available for biodegradability aspects there is still much scope to choose what combinations to be applied for as the most effective pre-treatment procedures.

6.5.2 Dewaterability

The reduction in the volume of sludge mixed with wastewater is a very crucial task. The sludge dewatering characteristics can be altered by pre-treatment processes like; thermal methods, ultra-sonication, electrolysis, microwave assisted irradiation, Fenton's reagent, alkaline pre-treatment, electrocoagulation, microwave-acid pre-treatment, chemical (acid/alkali) pre-treatment, and gamma irradiation [4].

6.5.3 Renewable Energy Production

Sustainable energy sources are one of the key requirements of the modern generation. The depeleting fossil fuels tend to produce a push to more research in this area. The gaseous substitutes have been generated as a potentially promising alternative. To conquer the energy crisis and bring in the circular economy concept in managing such waste, future research should be planned accordingly. These alternative fuels usage will also control the green-house gas emissions which will make the surrounding healthier [17]. Biological processes like anaerobic digestion when carried out on specific waste streams can generate bio-fuels potentially, this enables both aims of environmental protection and pollution control in an economic way to be achieved [2].

6.5.3.1 Bio-gas Production

Bio-methane production can be done by utilizing the organic components present in the industrial wastewater. The reactions involved in its production are [2];

$$CH_3COOH \rightarrow CH_4 + CO_2$$

$$4H_2 + CO_2 \rightarrow CH_4 + 2H_2O$$

Its production is, however, hindered because of the presence of chromium and sulphide in the substrate [4]. The factors that usually govern the production rate of bio-methane are; concentration of organics in the substrate, other components of the substrate, pH, organic loading rate, hydraulic retention time, etc. [2]. Pre-treatment is henceforth suggested to ensure better quality substrate for bio-methane production which can easily reduce the usage of fossil fuels on a day to day basis. Pre-treatments like microwave, ultra-sonication and chemical-mechanical were carried out on pulp mill wastewater and significant increase in the quantity of bio-methane production was obtained by microwave treatment [4]. Bio-methane has great potential to be used as a main fuel in developing and industrialized countries where availability of organic content is extensively present.

6.5.3.2 Bio-hydrogen Production

Biohydrogen is a suitable imperishable source of energy. It has countless advantages over remaining fuels such as, high viscosity, green burning, high heating value, etc. Even from economic point of view biohydrogen production can be considered as cost effective process [2]. Typical wastewaters coming from the paper and pulp industry, textile industry, petrochemical industry, etc., can be treated by pre-treatment processes such as the biological method (MCE), activated carbon, etc., which will significantly support the production of bio-hydrogen. Maintaining organic loading rate (OLR) is the key factor in bio-hydrogen production. The higher the OLR, the greater is the bio-hydrogen production as there is more hydrolysis acidification in the substrate [2].There is still a lot of scope to study the mechanisms involved in bio-hydrogen production by the use of different kinds of wastewaters. The number of studies will then ensure its application on practical sites and will also give an idea about its potential as an alternative fuel in terms of economy creation [2,4].

6.5.3.3 Bio-ethanol Production

This fuel derives its raw material from industrial waste rather than food waste, which makes it unique as a desirable alternative for future use. This resourceful characteristic of bio-ethanol ensures its continuous demand as a fuel. Several pre-treatment technologies like mechanical, chemical, thermal, thermos-chemical, when employed in particular combinations shows significant increase in the bio-ethanol production when practiced on several industrial wastes [4]. Further study of these combinations can be carried forward to find the modest combination for ethanol generation.

6.5.3.4 Bio-diesel Production

This optional fuel is found to emit lower sulphur di-oxide into the atmosphere and hence is a greener alternative to fossil fuels. This fuel can be mixed with petrol in proven proportions and can be used by automobiles making them less dependent on petrol. This would also be feasible from an economic point of view [17]. The production of bio-diesel is being studied by a large number of researchers, but its affordability to apply the processes on industrial scale is still left untouched. For now, its benefits do not balance the production cost involved. The process also has drawbacks on its operating stages like harvesting process making dewaterability of the substrate difficult [17].

6.6 Pre-treatment of Types of Industrial Waste

6.6.1 Major Industries and their Pre-treatment Technologies

Almost all industries produce wastewater as an unwanted by-product. The major polluting industries producing huge amounts of wastewater are discussed in detail here. Moreover, these waste water uses huge amount of energy which are represented in Table 6.1.

6.6.1.1 Petrochemical Industry

The wastewater produced from the petrochemical industry has a heterogeneous composition and assorted physiochemical parameters (e.g., pH, conductivity, COD, etc.). There are many pre-treatment processes that can be employed to treat this specific discharge like; ozonation, coagulation, photocatalytic degradation, filtration and adsorption [4]. Coagulation and flocculation methods are employed traditionally and are found to reduce the toxic concentration of substrates. Anaerobic and aerobic processes can also be employed to treat the biological components of the substrate. All these treatments are carried out to reduce the level of membrane fouling which is mainly caused by organic and inorganic materials. Mostly for secondary treatment of this sludge reverse osmosis is preferred. So, the pre-treatment can be carried out by various techniques like granulated activated carbon filtration (GAC), ultrafiltration (UF), nano filtration (NF), combination of both GAC+NF, etc. Combining UF with coagulation and flocculation provided an end product having lower turbidity, SS and COD levels [4]. AOP is one of the most adopted techniques as a pre-treatment. A few other pre-treatment technologies that are photocatalytic degradation, thermolysis, ozonation or their combination, etc., have also been studied by different researchers [4]. Pre-treatment given to the wastewater from this industry supports methane formation which can be a good alternative for traditional fuels.

6.6.1.2 Paper and Pulp Industry

The paper and pulp industries use natural resources on a large scale. The consumption of a high volume of water also results in the production of a high quantity of wastewater. The wastewater is comprised of lignin, polyphenols, cyanide, chlorinated organic compounds, aromatic compounds, etc. The discharges from the paper and pulp industry is fatal to aquatic life. The disruption in carbohydrate cells, ion effect and biotransformation activities of reagents are visibly affected by this wastewater [4]. This effluent cannot be simply treated by biodegradation techniques due to its complex composition. Also as the process is costly and even the treated effluent cannot reach the desired standards. Several studies proposed the use of a few pre-treatment technologies, i.e., thermochemical precipitation, advanced oxidation processes, adsorption, electrocoagulation and enzymatic pre-treatment which are believed to produce desirable results [4]. The only drawback is that most studies were carried out in laboratories and only a few were tested as pilots due to a lack of economic resources. Electrocoagulation was carried out as a pre-treatment followed by an UV/oxidant system which gave good results in the laboratory. Another experiment in which permanganate (PM) oxidation/coagulation was chosen as a pre-treatment technique for secondary treatment by the electro-Fenton (EF) process and a subsequent sulfate radical (SR)-based process (Co3O4/ per oxy mono sulfate /UV) gave good results [4]. The pre-treated wastewater from the paper industry can be reutilized in bio-methane production [4]. But the production is very low compared to the cost of treatments applied and hence more feasible measures are desirable for the treatment of this wastewater [18].

6.6.1.3 Coal Manufacturing Industry

Fossil fuels are non-renewable resources, although cost effective they generate several environmental concerns. Acquisition of fossil fuels like coal involves regressive procedures and thus impacts the environment severely. The need to manage this issue has imposed a barrier to the use of pre- and post-treatment technologies to ensure less harm to the environment. The effluent that needs treatment is mainly fly ash, when mixed with toxic heavy metals it needs to be prevented from going into landfills, which can later form leachate and pollute groundwater directly. Fly ash has to be controlled to to stop it from moving with wind on land and scattering over undesirable areas. Pre-treatment of fly ash can change the physical and chemical characteristics of it which makes it available for different use. The pre-treatment of fly ash by hydrogen-peroxide and carbonation is studied by researchers to protect the inertness of fly ash [4]. The common contaminants from coal industry like ammonia, cyanides, phenolics and polycyclic aromatic hydrocarbons can be functionally removed by advanced oxidation technologies. The wastewater produced from coal gasification units are also given pre-treatment with acid/alkali in the industries to make it usable for other purposes. The treatment involving Fenton oxidation (HFO) with an anoxic moving bed biofilm reactor (ANMBBR) and biological aerated filter (BAF) along with FeOx/SBAC as the catalyst is approved as a profuse combination of technologies [4]. The catalytic ozonation method using nano-MgO when carried out on CGW shows significant results [4]. Many pre-treatment processes are found to reduce the high levels of fly ash and enables its usage for various purposes. The adoption of laboratory theories on pilot or large industrial scale will definitely ensure environmental safety and will also generate economy.

6.6.1.4 Textile Industry

Textile industries require a lot of chemicals and dyes for the production of the desired products. Due to this extensive use the discharge produced is difficult to identify on a qualitative basis and is also difficult to treat. Advanced chemical treatments are proposed in different studies to treat the wastewater comprising of dyes. These techniques are namely; electrochemical treatment, coagulation and flocculation, photocatalytic oxidation, and adsorption [4]. The decolourization of textile effluents is a pivotal task and this can be carried out effectively by advanced oxidation processes (AOP). The procurement of already used important elements from this wastewater is made easy accordingly, which makes the overall process cost effect and beneficial [4]. Ozonation is carried out on this substrate which reduces its toxicity at large scale but the exposure time is a bit longer which eventually makes this technique costly. The technology involving the combination of Al2(SO4)3 and a cationic organic flocculant resulted in the reduced levels of total organic carbon (TOC), COD, AOX, BOD and anionic surfactants. This method could also decolourize the substrate effectively and hence proved worthy [4]. Biological treatment using a variety of fungi has also been studied widely which resulted in fruitful results. Pre-treated discharge from this industry has found its way into the construction industry as it can produce good compressive strength building blocks [4].

6.6.1.5 Polyester Industry

The production of polyester fibre is necessary for the works of the textile industry. The involvement of various surfactants, additives and detergents, including wetting agents, softening agents, antioxidant, and antiseptics pollutes the water discharged from this industry [19]. The pollutants are toxic in nature, cannot be degraded easily and assist in the formation of toxic foam on the surface of the wastewater. Additives involved in improving the quality of the polyester products also contributes to the acidic foam formation which makes the task of cleaning more difficult [19].The wastewater produced from this industry is difficult to treat by classical methodologies and so the application of advanced systems is mostly used to gain quality results. To treat this waste stream, as a preliminary treatment interior micro-electrolysis is used followed by anaerobic digestion. The major reactions maintaining the process can be listed as redox, flocculation and electrophoresis. This pre-treatment technique is chosen as it operates at low cost, no extra expenses is needed to buy coagulants and even no extra energy is needed for its operation [19].Micro-electrolysis is found to effectively remove surfactant reagents which are the major causes of foaming in the substrate. Biodegradability of the organics is increased; the BOD/COD ratio also shows positive change; several non-biodegradable compounds were also found missing in the end product.

6.6.1.6 Pharmaceutical Industry

Pharmaceutical industries generate varieties of toxic waste, harmful contaminants throughout their production and disposal treatments. Incineration is practiced on a large scale to dispose of the waste from this industry which also produces harmful gases which readily pollute the surrounding air. For this study though we will just focus on liquid waste. Antibiotic production wastewaters are difficult to treat due their antibacterial potential, so before their treatment biologically, advanced procedures are needed [20]. The pre-treatment processes to handle such wastewater are Fenton oxidation, wet-air oxidation, coagulation/flocculation and filtration. The electro-Fenton process could effectively break down phenol, pyridine and quinolone which renders maximum toxicity in the substrate. Ozonation proves to be a great technique to handle waste from this industry. The composition of pharmaceutical waste is hard to study. It comprises of high concentrations of organic matter, microbial toxicity, high salt content which do not degrade easily and are also difficult to treat. Only with the proper treatment can the pharmaceutical waste discharge reach the disposal standards. Further study is required to treat this specific type of wastewater.

6.6.1.7 Olive Oil Industry

Olive oil production is carried out seasonally around the globe. The wastewater produced from this industry constitutes of very high organic load, recalcitrant in nature and a high amount of toxicity/phyto-toxicity in its components [9].

Biological treatment is usually carried out to treat this stream. However, application of AOP's is preferred to get better results. The main issue with the wastewater is the handling of high levels of COD and BOD in the waste stream. Factors which affect the characteristics of water from this industry are mainly the method of extraction, type and maturity of olives, region of origin, climatic conditions and associated cultivation/processing methods [9]. The challenge for the treatments applied is the reduction of toxicity levels which are extremely harmful for the environment. Because of the seasonal production of olive oil, it is hard to keep the treatment plant working throughout. And so such measures are required which are cost effective and eco-friendly at the same time. Olive farming is carried out in a scattered manner by small farmers, and the production of oil too is done as a small family business. So the treatment of the wastewater produced is less affordable for such enterprises. For treating OMW hybrid treatment technology is preferred. All the treatments such as physical, chemical and biological are applied together to obtain the desired results. The most common physical treatments are dilution, evaporation, sedimentation, filtration and centrifugation. The feed after this treatment is not of sufficient standard and so is sent for post-treatments [9]. Centrifugation and filtration when applied on the substrate causes phase separation and exclusion of organic matter. Thus making it easier for the materials to become more conductive in increased pH and get removed with ease [9]. Adsorption or desorption requires a lot of attention as after a certain contact period, phenols and organics present in the wastewater initiates desorption causing the failure of the process [9].

Thermal pre-treatments are given to the wastewater from this industry. Combustion and pyrolysis is carried out using fossil fuels and the exhaust gases are exposed to atmosphere polluting it on large scale. Henceforth making the process costly and polluting. Electro-coagulation is also carried out for this stream, it is found to reduce the quantity of overall sludge produced at the end of the final treatment. It is also applied because of its efficient COD removal quality, and also from an economic point of view [9]. Biological measures are quite famous for this industry as there is huge involvement of organics in the wastewater. Aerobic digestion and anaerobic digestion when carried out conventionally do not produce high quality results as compared to when the feed is pre-treated by various methods. Physio-chemical pre-treatments such as oxidation, coagulation and extraction can effectively reduce the polyphenol content. The feed when it is then applied for further treatment reduces COD and oleic acid from the end product [9]. Treatment with ozone, powdered activated carbon, wet air oxidation are also some fruitful pre-treatments. Ozone treatment is used when it is desired to break aromatic and double bonds of the components present. Fenton treatment accompanied by UV or solar light treatment, although costly, produces end products matching the standards specified. There are certain studies suggesting the use of aerobic treatment prior to anaerobic treatment. This combination was specified

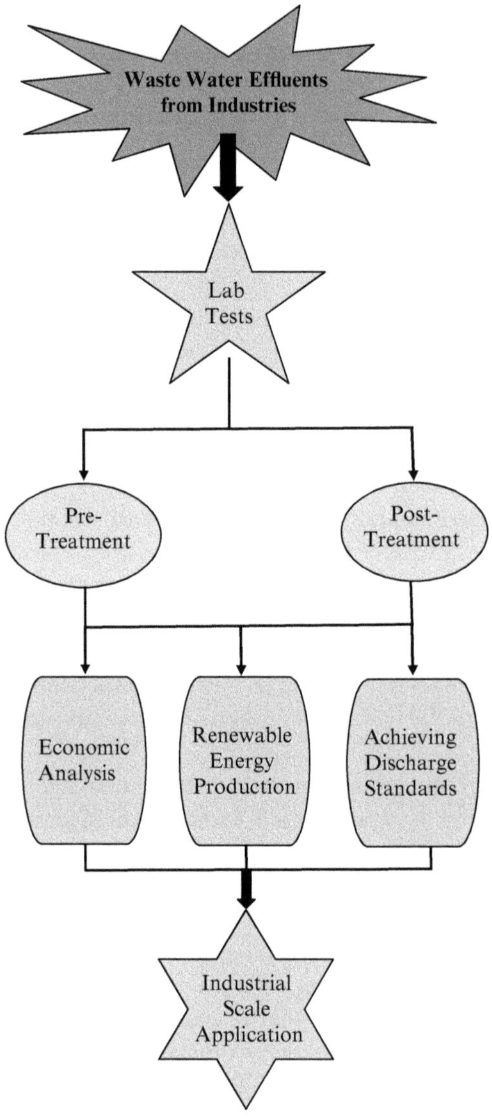

Figure 6.10 Outline for incorporating advanced treatments to wastewater from industries.

as the first reactor being an up-flow anaerobic reactor and the second reactor being a fixed bed reactor. This combination efficiently reduced the COD level [9]. Composting is carried out on organic matter to convert them into plant friendly fertilizers. OMW can be a great choice to obtain high quality compost. The compost acquired from this substrate is rich in nutrients for plant growth, is more stable in its form and also has no heavy metal content [9]. All these treatment schemes are applied in order to achieve the target of sustainability and circular economy from the perspective of energy recovery and other compounds like phenols which can be used for other processes.

6.6.1.8 Fertilizer Industry

Triazine is a nitrogen compound used in the manufacturing of pesticides for farm use. The equipment when washed or when water is mixed with the pesticides residues or content, renders toxicity to it. A few classical technologies used for treating such wastewater can be listed as; hydrolysis, high-temperature incineration, biological processes, deep well injection, chemical precipitation, chemical oxidation, activated carbon adsorption, resin adsorption, steam or air stripping, etc. Nowadays, advanced technologies such as; photocatalytic oxidation, photo-assisted Fenton reaction, ultrasonic radiation are brought into practice [21]. Internal micro-electrolysis is a fresh technique which works in a similar manner to electro-chemical techniques. IME is used as a pre-treatment here to treat this industries wastewater. The subsequent treatment given after is mostly biological, as the feed from the first treatment supports the degradation of toxic organics in wastewater. It is applied at large scale as it can effectively reduce odour and colour, also can lower the suspended solids concentration, COD in the final discharge [21]. Acrylonitrile–butadiene–styrene (ABS) resin production wastewater is treated by this method [22]. The results achieved while tackling concentrated toxic waste were so good that now IME can also be applied to other industrial wastewater having similar components. It is a comparatively cheap way and is also eco-friendly, hence its use is promoted [21].

6.6.2 Cost Analysis

Table 6.1 Energy Usage of few water and water sources [14].

Feed	Conv. act sludge (kW h m^{-3})	Pre-treatment (kW h m^{-3})	RO system (kW h m^{-3})	Total treatment (kW h m^{-3})
Surface water				0.15–0.3
Wastewater	0.3–0.6	0.1–0.2	0.4–0.5	0.8–1.3
Wastewater MBR		0.8–1.0	0.4–0.5	1.2–1.5
Brackish (930–2200 ppm)		0.1–0.3	0.6–0.9	0.8–1.0
Brackish – Tidal Estuary		0.29	1.38	1.67
Seawater (medium salinity)		0.3–1.0	2.0–3.0	2.3–4.0

6.7 Future Prospects

Maximum studies on pre-treatment technologies are carried out at laboratory scale successfully. Integration and intensification in the adoption of these theories practically at industrial scale requires a further push. Further studies need to be carried out on the by-products formed after the pre-treatment processes to ensure good quality feed for the further final treatment. It is necessary that the entire treatment unit setup is cost efficient and gives good results matching the fixed standards of discharge, which requires further research. Usually all the waste coming out of different industries is collected and transported to a city's wastewater treatment facility, so many studies suggest the use of a comprehensive treatment unit comprising of all the technologies which can treat this waste altogether, the outline of this is shown in Figure 7.4. If such a treatment unit is brought into existence, then the power production from different fuels like bio-methane and bio-hydrogen can be increased to a greater extent. Micro-pollutants coupled with heavy metals are found in many substrates from different industries and require a lot of study to formulate methods to treat them separately in an efficient way.

6.8 Conclusion

The introduction of pre-treatment technologies for treating industrial wastewater have numerous benefits. It not only assures reduced usage of resources but also keeps a check on proper disposal of the end products. Further optimization of these treatment technologies requires a lot of attention to bring these theories from books to practice. Certain disparities can come across while implementing some pre-treatment technologies so it is important to carry out the pre-requisite studies of the substrate characteristics and also its cost effectiveness as an overall process. Keeping in mind the sustainability concept, zero liquid discharge should be the ultimate aim. And hence to ensure that a battery of tests should be carried out in different combinations prior to the actual pilot or industrial scale implementation. To control the rising environmental concerns more research and analysis needs to be carried out on major industries. The effective disposal of their effluents will immediately show a positive change in the surrounding animal and human health and living conditions.

References

1 Rozkošný, M., Kriška, M., Šálek, J. et al. (2014). Natural technologies of wastewater treatment. ISBN: 978–80-214–4831-5.

2 Tian, X., Song, Y., Shen, Z. et al.(2019). A comprehensive review on toxic petrochemical wastewater pretreatment and advanced treatment. *Journal of Cleaner Production PII* S0959–6526 (19): 33562. doi: 10.1016/j.jclepro.2019.118692.

3 Vincent-Vela, M.C., Álvarez-Blanco, S., Lora-García, J. et al. (2015). Application of several pretreatment technologies to a wastewater effluent of a petrochemical industry finally treated with reverse osmosis. doi: 10.1080/19443994.2014.939866.

4 Prabakar, D., Subha, S.K., Manimudi, V.T. et al. (2018). Pretreatment technologies for industrial effluents: critical review on bioenergy production and environmental concerns. *Journal of Environmental Management* 218 (2018): 165e180. doi: 10.1016/j.jenvman.2018.03.136.

5 Nascimento, T.A., Fdz-Polanco, F., and Peña, M. (2018). Membrane-based technologies for the up-concentration of municipal wastewater: a review of pretreatment intensification. doi: 10.1080/15422119.2018.1481089.

6 Semblante, G.U., Lee, J.Z., Lee, L.Y. et al. (2018). Brine pre-treatment technologies for zero liquid discharge systems. *Desalination* 441 (2018): 96–111. doi: 10.1016/j.desal.2018.04.006.

7 Anjum, M., Al-Makishah, N.H., and Barakat, M.A. (2016). Wastewater sludge stabilization using pre-treatment methods. *Process Safety and Environment Protection* S0957–5820 (16): 30081–30087. doi: 10.1016/j.psep.2016.05.022.

8 Wang, Q., Wei, W., Gong, Y. et al. (2017). Technologies for reducing sludge production in wastewater treatment plants: state of the art. *Science of the Total Environment* xxx (2017): xxx–xxx. doi: 10.1016/j.scitotenv.2017.02.203.

9 Paraskeva, P. and Diamadopoulos, E. (2006). Technologies for olive mill wastewater (OMW) treatment: a review. *Journal of Chemical Technology and Biotechnology* 81 (2006): 1475–1485. doi: 10.1002/jctb.1553.

10 Sharmila, V.G., Banu, J.R., Kim, S.-H., and Kumar, G.(2020). A review on evaluation of applied pretreatment methods of wastewater towards sustainable H2 generation: energy efficiency analysis. *International Journal of Hydrogen Energy* xxx (x x x x): xxx. doi: 10.1016/j.ijhydene.2020.01.081.

11 Durham, B., Marguerite Bourbigot, M., and Pankratz, T. (2001). Membranes as pretreatment to desalination in wastewater reuse: operating experience in the municipal and industrial sectors. *Desalination* 138 (2001): 83–90. doi: 10.1016/S0011-9164(01)00248.

12 Kim, S.L., Paul Chen, J., and Ting, Y.P. (2002). Study on feed pretreatment for membrane filtration of secondary effluent. *Separation and Purification Technology* 29 (2002): 171–179. doi: 10.1016/S1383-5866(02)00073-4.

13 Xu, Y., Li, Y., and Hou, Y. (2018). Reducing ultrafiltration membrane fouling during recycled paper mill wastewater treatment using pretreatment technologies: a comparison between Coagulation and Fenton. *Journal of Chemical Technology and Biotechnology* doi: 10.1002/jctb.5826.

14 Pearce, G.K. (2008). UF/MF pre-treatment to RO in seawater and wastewater reuse applications: a comparison of energy costs. *Desalination* 222 (2008): 66–73. doi: 10.1016/j.desal.2007.05.029.

15 ÁLvarez, J.A., Rúız, I., and Soto, M. (2008). Anaerobic digesters as a pretreatment for constructed wetlands. *Ecological Engineering* 33 (2008): 54–67. doi: 10.1016/j.ecoleng.2008.02.001.

16 Tejedor-Sanz, S., Ortiz, J.M., and Esteve-nú Ñez, A.(2017). Merging microbial electrochemical systems with electrocoagulation pretreatment for achieving a complete treatment of brewery wastewater Chemical. *Engineering Journal* S1385–8947 (17): 31390–31396. doi: 10.1016/j.cej.2017.08.049.

17 Kadir, W.N.A., Lama, M.K., Uemura, Y. et al. (2018). Harvesting and pre-treatment of microalgae cultivated in wastewater for biodiesel production: a review. *Energy Conversion and Management* 171 (2018): 1416–1429. doi: 10.1016/j.enconman.2018.06.074.

18 Saha, M., Eskicioglu, C., and Marin, J. (2011). Microwave, ultrasonic and chemo-mechanical pretreatments for enhancing methane potential of pulp mill wastewater treatment sludge. *Bioresource Technology* 102 (2011): 7815–7826. doi: 10.1016/j.biortech.2011.06.053.

19 Yang, X. (2009). Interior microelectrolysis oxidation of polyester wastewater and its treatment technology. *Journal of Hazardous Materials* 169 (2009): 480–485. doi: 10.1016/j.jhazmat.2009.03.123.

20 Yi, Q., Gao, Y., Zhang, H. et al. (2016). Establishment of a pretreatment method for tetracycline production wastewater using enhanced hydrolysis. *Chemical Engineering Journal* 300 (2016): 139–145. doi: 10.1016/j.cej.2016.04.120.

21 Cheng, H., Weipu, X., Liu, J. et al. (2007). Pretreatment of wastewater from triazine manufacturing by coagulation, electrolysis, and internal microelectrolysis. *Journal of Hazardous Materials* 146 (2007): 385–392. doi: 10.1016/j.jhazmat.2006.12.038.

22 Lai, B., Zhoua, Y., Qina, H. et al. (2012). Pretreatment of wastewater from acrylonitrile–butadiene–styrene (ABS) resin manufacturing by microelectrolysis chemical. *Engineering Journal* 179 (2012): 1–7. doi: 10.1016/j.cej.2010.12.089.

Index

Application of Sewage Sludge in Industrial Wastewater Treatment, First Edition. Edited by Maulin P. Shah.

www.ingramcontent.com/pod-product-compliance
Ingram Content Group UK Ltd.
Pitfield, Milton Keynes, MK11 3LW, UK
UKHW020208290926
15470UKWH00001B/62

9 781119 857365